# CONFIDENTIALITY OF DATA AND CHEMICALS CONTROL

ORGANISATION FOR ECONOMIC CO-OPERATION AND DEVELOPMENT

The Organisation for Economic Co-operation and Development (OECD) was set up under a Convention signed in Paris on 14th December 1960, which provides that the OECD shall promote policies designed:

- to achieve the highest sustainable economic growth and employment and a rising standard of living in Member countries, while maintaining financial stability, and thus to contribute to the development of the world economy;
- to contribute to sound economic expansion in Member as well as non-member countries in the process of economic development;
- to contribute to the expansion of world trade on a multilateral, non-discriminatory basis in accordance with international obligations.

The Members of OECD are Australia, Austria, Belgium, Canada, Denmark, Finland, France, the Federal Republic of Germany, Greece, Iceland, Ireland, Italy, Japan, Luxembourg, the Netherlands, New Zealand, Norway, Portugal, Spain, Sweden, Switzerland, Turkey, the United Kingdom and the United States.

Publié en français sous le titre

LE CARACTÈRE CONFIDENTIEL DES DONNÉES
ET LE CONTROLE DES PRODUITS CHIMIQUES

.°.

FOREWORD

The laws and administrative practices established in OECD Member countries to control chemicals generally require the transfer of data from industry to governments. Data and information about the potential hazards of chemicals is a prerequisite for the effective protection of human health and the environment.

The authorities responsible for chemicals may receive data and information from sources other than directly from industry: information may be received from other departments within a national administration or from a different national administration. In all these cases there may be a need to treat the data in a confidential manner; for example because of the need to protect business secrets or the proprietary rights to data, or because this is an established feature of government-industry relations in a country.

However, the national authorities may be required to provide for public participation in their decision making process, or to make the bases of their decisions publicly available. Such requirements, involving the disclosure of information, may conflict with the need to treat certain data in a confidential manner.

The way this dilemma is solved has international implications. Different policies for handling sensitive or confidential information may have the potential to effect trade in chemicals, particularly when chemicals control laws require the submission of such information to government. Also, human health and environmental protection may benefit from the exchange of information between governments; but exchanges of some information require a common understanding on the identification and treatment of confidential data.

These issues have been addressed by an OECD group of experts, led by France, and this book summarises the work which they carried out and the conclusions they reached. This work is part of a broader programme of work in OECD which addresses international aspects of chemicals policy. Apart from Confidentiality of Data, the first phase of the Special Programme on the Control of Chemicals included work on an International Glossary of Key Terms, Good Laboratory Practice and Information Exchange.

The first part of the book is a factual review of legal provisions and administrative practices in Member countries related to the confidentiality of data, specifically the confidentiality of data on chemicals. The information was collected by a questionnaire circulated in 1979 and updated to reflect the situation as it was in 1981. It will interest industrial submitters of data to national administrations and those who receive data at the level of government or administration, as well as scholars of comparative law and administration.

The report of the group of experts constitutes the second part of the book. It reflects on the various aspects of confidentiality, and areas are identified where international cooperation might be of benefit. Proposals are made to facilitate wider dissemination of data on the basis of a list of non-confidential data; principles are suggested which enable the transfer between governments of confidential information; and a principle is proposed for the protection of proprietary rights to data on new chemicals.

This document has been derestricted under the authority of the Secretary-General. The views expressed do not necessarily reflect those of OECD or its Member countries; they provide the basis for future consideration of these issues by the OECD Council and its subsidiary bodies.

Jim MacNeill
Director
OECD Environment Directorate

*Also available*

**CHEMICALS CONTROL LEGISLATION: AN INTERNATIONAL GLOS-SARY OF KEY TERMS (1982)**
(59 82 02 1) ISBN 92-64-12364-4                           £6.80     US$13.50     F68.00

**GOOD LABORATORY PRACTICE IN THE TESTING OF CHEMICALS (1982)**
(59 82 04 1) ISBN 92-64-12367-9                           £6.80     US$7.00     F35.00

**OECD GUIDELINES FOR TESTING OF CHEMICALS (July 1981)**
(97 81 05 1) ISBN 92-64-12221-4,   734 pages              £35.00    US$80.00    F360.00

**CONTROL OF CHEMICALS IN IMPORTING COUNTRIES (August 1982)**
(59 82 01 1) ISBN 92-64-12272-9,   196 pages              £5.40     US$12.00     F54.00

**OECD AND CHEMICALS CONTROL (March 1981)**
(02 81 03 1) "OECD Background Paper"                       £4.00     US$9.00      F40.00

*Prices charged at the OECD Publications Office.*

*THE OECD CATALOGUE OF PUBLICATIONS and supplements will be sent free of charge on request addressed either to OECD Publications Office, 2, rue André-Pascal, 75775 PARIS CEDEX 16, or to the OECD Sales Agent in your country.*

# TABLE OF CONTENTS

PART II:  CONFIDENTIALITY OF DATA AND PROTECTION
OF PROPRIETARY RIGHTS IN RELATION
TO THE DISCLOSURE AND THE EXCHANGE
OF INFORMATION ON CHEMICALS

(Final Report of the Group of Experts on
Confidentiality of Data)

## PREFACE BY THE CHAIRMAN
## OF THE GROUP OF EXPERTS

The confidential nature of certain data relating to chemicals, although in fact fully justified on economic and commercial grounds, is an obstacle to the dissemination of information among those concerned in various ways with the protection of man and the environment. It is in particular an obstacle to the transfrontier exchange of information between the national authorities responsible for the implementation of chemicals control, a new chapter of environmental policy which consists essentially in ensuring that before being put into circulation and produced on a large scale, chemicals are fully evaluated with regard to their effects on human health and the environment.

In an attempt to remove this obstacle as far as possible, a group of experts, led by France, was established under the OECD Special Programme on the Control of Chemicals to look into the problem of confidentiality of data.

Before contemplating any changes in the present situation, however, it was first necessary to find out what the situation in the OECD countries was as regards the protection and disclosure of confidential information and to analyse its various aspects.

The first part of this publication sets out the results of that study. It contains a wealth of information which has been extremely useful to the Group of Experts in its work and I wish to express my thanks to Mrs. Persoz of the French delegation, to the group of experts, and to the members of the Secretariat of the OECD who contributed to the survey of national practices.

The reader will surely be struck by the OECD countries' wide variety of approach, sometimes deeply rooted in national ways of thought, sometimes based on the sources of national law, and sometimes simply a matter of the varying degrees of progress made by countries in implementing new legislation on chemicals control.

The differences in national attitudes with respect to confidentiality and disclosure of data are but one of the factors which, while complicating the task set aside for the groups of experts, has made the exercise very rewarding.

The final report of the groups which constitutes the second part of this publication reflects
the constructive spirit which guided the group
throughout its work and which was one of conciliation
and compromise between national sensitivities and
interests of the many parties concerned by the issue
of confidentiality.

The group in presenting its report is well
aware of its limitations and does not pretend to bring
definitive answers to the many questions that can be
raised. The group certainly went as far as it could
in presenting concrete proposals in three domains: the
acceptance of the non-confidential nature of certain
data necessary for hazard assessment, the pro- tection
of proprietary rights of data and the exchange of
confidential information between the authorities of
Member countries.

Much remains to be done in order to achieve,
not necessarily uniformity - hardly desirable since
many of our differences are mutually enriching - but
the indispendable harmonization of certain provisions
which will enable the public to be better informed and
protect at the same time legitimate commercial
interests. This harmonization is the goal to which
our efforts must be bent in the years ahead.

I believe that the present publication will be
useful to everyone concerned by these problems, particularly those who in the future carry on with the
enormous task that the group has merely begun.


J.P. Parenteau
Inspecteur Général de l'Environnement
au Ministère Français de l'Environnement

PART I


CONFIDENTIALITY OF DATA AND INFORMATION DISCLOSURE
UNDER CHEMICALS CONTROL PROVISIONS IN
OECD MEMBER COUNTRIES

## SUMMARY

Soon after its formation in 1979, the Group of Experts on the Confidentiality of Data circulated a questionnaire concerning policies and regulations on confidentiality and disclosure of information. The questionnaire focussed on data necessary or useful for hazard assessment of chemicals and other purposes related to the protection of man and the environment. Australia, Austria, Belgium, Canada, Denmark, the Federal Republic of Germany, Finland, France, Italy, Japan, the Netherlands, New Zealand, Norway, Sweden, Switzerland, the United Kingdom, the United States of America and the Commission of the European Communities responded to the questionnaire.

This report presents on a country-by-country basis the material contained in the responses arranged under several chapters. The information in the report applies to policies and regulations existant in early 1981.

Chapter 1 of this report examines the way governments and administrations in OECD Member countries treat the confidentiality of documents and information they possess. Section 1.1 analyses their perception and definition of the concept of "confidentiality" in general and trade secrets or commercial information in particular. A related issue, the public's right to obtain official documents, is discussed in Section 1.2, since public access can be another way of releasing official material. In addition, exemptions from disclosure under public access rules, based on the concept of trade secrets and commercial confidentiality, are reviewed.

Chapter 2 discusses the confidentiality issue in terms of chemical control requirements in Member countries. Section 2.1 outlines the type of data required under notification programmes and authorisation procedures. Section 2.2 on the handling of confidential information under the chemical laws constitutes the core of this report. It discusses the treatment given to data on chemicals contained in dossiers submitted to administrations, as well as the policies of determining and protecting confidential material.

Chapter 3 concerns the official transmission of confidential data. Section 3.1 discusses the exchange of information between government officials within a country. Section 3.2, which covers international exchanges, is brief because most national laws have not yet addressed the issue.

# CHAPTER 1

## CONFIDENTIALITY

1.1:     GENERAL STATUTES PROTECTING CONFIDENTIAL
         INFORMATION HELD BY NATIONAL ADMINISTRATIONS

In general, the need to protect commercially valuable data submitted to governments is recognised in all OECD Member countries. This section explores primarily how the concept of commercial and industrial confidentiality appears in various national statutes which apply to civil servants and the official documents that go through their hands. However, some statutes do apply to private employees and the confidential information to which they are privy.

In some Member countries with a common law tradition, the United Kingdom, Australia, and the United States, there is a legal remedy whereby the supplier can sue under civil tort law for "breach of confidence". This right, which exists separately from the protection granted by legislation, is derived from common law. Elements of the tort are set out in the sections below on Australia and the United Kingdom. The tort exists in some states in the US, but is not yet applicable on the federal level.

### AUSTRALIA

Australia has a federal system of government under which the Commonwealth, or national, administration is responsible for certain areas of regulation. Other areas are left to the individual State administrations. There is no legislation at the federal level dealing specifically with confidentiality of data. However, a Commonwealth official may be prosecuted under Section 70 of the Crimes Act and Public Service Regulations 34 and 35 for disclosing any information marked "In Confidence (Commercial)". On the State level, there are no general codes or statutes concerning industrial or commercial confidentiality. However, restraints similar to those on Commonwealth officials are also in effect for State government officials.

Breach of Confidence: the civil remedy of breach of confidence is also available to protect confidentiality. The courts can award money damages and/or an injunction against further disclosure provided that

(i) the data have the necessary "quality of confidentiality about them";

(ii) the data were given in return for some obligation to keep them secret; and

(iii) the data have been used without the supplier's authorisation, to the detriment of the supplier (*).

## BELGIUM

The concept of confidentiality in Belgium is inherent in general regulations requiring official secrecy. In addition to a general provision for professional secrecy in Article 458 of the Penal Code, there is a specific obligation for civil servants not to disclose confidential information. The Arrêté Royal of 2nd October, 1937, prohibits civil servants, whether active or retired, from disclosing facts acquired in exercising their functions which have the character of secrecy due to their nature or by rule. Violations of this regulation are sanctioned by disciplinary measures. Commercial and business secrets are not defined in Belgian law and are left for the courts to determine in disputed cases.

## CANADA

All public servants come under the jurisdiction of the Official Secrets Act which prohibits them from releasing any information received in confidence unless specifically instructed to do otherwise.

Industrial and commercial data are collected under various laws, some of which contain provisions for confidentiality. These, in as far as they touch upon control of chemicals, are considered in Chapter 2.

---

(*) Malone v. Metropolitan Police Commission, 2 W.L.R. 700, 728 (1979) (L. Megarty) (quoting) Saltman Engineering Co. Ltd. v. Campbell Engineering Co. Ltd.

## DENMARK

The concept of professional secrecy, as it appears in Danish law, protects confidential information submitted to the government.

The _Civil Penal Code_ (Section 152) forbids civil servants from disclosing data acquired in the course of their work. The law identifies the protected information as data which are either treated as confidential or defined as confidential by a statute or other regulation. Data treated as confidential can include information which may not be protected by other specific legal provisions or regulations. For example, it would apply to those cases where a public authority collects data from a citizen who is under no obligation to submit them, but does so merely with the understanding that the data will remain confidential.

The _Public Servants Act_ (Section 10, para. 2) requires officials to maintain, both during and after government service, the secrecy of information they might have acquired through their work.

As noted in Section 1.2, provisions of these laws do not prevent persons from obtaining data from ad- ministrations. Civil servants who release material under the public access law do not violate the pro- visions of the above laws.

## EUROPEAN COMMUNITIES

There are no pertinent general provisions for confidentiality in the European Communities. The provisions for handling confidential information in the notification dossiers required by the EC Council Directive 79/831 EEC are discussed in Section 2.2.

## FEDERAL REPUBLIC OF GERMANY

Paragraph 30 of the _Administrative Procedure Act_ requires that industrial, commercial and other confidential data in the government's possession remain secret. The term "Betriebs-und Geschäftsgeheimnisse" (enterprise and business secrets) is not defined in the Act and is left to judicial interpretation. Case law to date has only provided some general statements that data must not be disclosed because of industrial, commercial or other legitimate interests.

Before releasing data the government must obtain the consent of the submitter of the data, unless there is some other legal authorisation for disclosure. An action may be brought by the submitter before the administrative courts to prevent the wrongful disclosure of the data. Once confidential data have been released illegally, the supplier may obtain damages in a civil suit against the administration under the Federal Act concerning the Liability of the State (Staatshaftungsgesetz).

## FINLAND

The concept of confidentiality in Finland exists within the context of the Publicity of Official Documents Act (38/51) (discussed more fully in Section 1.2). The various classifications of government documents, as well as exceptions to the right of public access, are the means of protecting information submitted to administrations.

The classification of official documents in Finland determines the degree of protection accorded commercial data. According to the present interpretation of the Publicity of Official Documents Act (83/51), information in an administration's possession may belong to one of three categories: public documents, non-public documents, or confidential documents. Public documents are available to the public under the access law. Nonpublic documents include documents under preparation which become public as soon as the drafting and preparation are completed. Internal office proposals, such as drafts, reports, etc., do not become public upon completion. Confidential or secret documents generally may not be released by government officials. This last category includes documents kept secret to protect business interests. The regulations implementing the public access law (discussed in Section 1.2) provide the authorisation for withholding trade secrets.

In the event that confidential information has been wrongfully disclosed by an administration, the submitter of the data can seek redress by bringing an action under the Compensation Act. Under Section 19, Chapter 40 of the Penal Code, officials may be punished for improperly discharging their duties, such as wrongfully releasing trade secrets.

## FRANCE

French civil servants are bound by the general obligation of professional secrecy laid down in Article 378 of the Penal Code. Information held by the administration is further protected from disclosure by a rule in Article 10 of the Ordinance concerning the statute of civil servants of 4th February, 1959. In a definition issued by the Council of State on 6th February 1951, information requiring secrecy is that which is acquired by civil servants in the fulfillment of their duties containing facts that are confidential in nature or that have been submitted under the seal of secrecy.

The concept of industrial and business secrets has been introduced more recently by law No. 78-753 of 17th July, 1978 as a basis for denying access to administrative documents. However, no explicit definition is given; to date only the concept of manufacturing secrets (secrets de fabrique) has appeared in French jurisprudence. The term exists in Article 418 of the Penal Code which prohibits employees of an enterprise from disclosing manufacturing processes known to them.

## ITALY

Provisions for the protection of confidential information appear in the Italian Penal and Civil Codes. Professional secrecy in general is covered in Article 623 of the Penal Code. The Code punishes anyone who, because of his function, office or profession, has access to data intended to be kept secret concerning discoveries, scientific inventions or industrial applications and discloses the data or uses them to his own profit or to the profit of others.

Art. 326 of the Penal Code more specifically prohibits public officials or persons in charge of a public service from revealing administrative secrets. The notion of administrative secrecy may apply to information specified by law or statutory order or simply because of the nature of the information itself.

Article 2105 of the Civil Code prevents employees from divulging information about an enterprise's organisation or production methods or using them to the detriment of the enterprise. Recourse to the civil or criminal magistrate is available to persons whose interests have been compromised by the disclosure of confidential information.

## JAPAN

A Japanese public official must not divulge secrets acquired in his work at the risk of being fined or imprisoned. This pertains to secrets contained in all types of official documents and information (National Public Service Law, Articles 100, 109).

## NETHERLANDS

In addition to the exceptions to the public access law discussed in Section 1.2, confidential information is protected in the Netherlands. Under the Administration of Justice Act (the "AROB Act"), an action can be brought against the government for wrongful disclosure of confidential information in an administrative order. More generally, under the Central Government Personnel Act and the General Civil Service Regulations submitters of information may take action against officials who do not maintain the secrecy of government documents.

With regard to information disclosed by private employees, the Criminal Code (Article 272) prohibits divulging secrets obtained in the course of employment.

## NEW ZEALAND

The Official Secrets Act and its regulations, together with the State Service Act, provide protection for any document or information held by government organisations. Disclosing government-held information for purposes other than those specified by the administration is illegal. The statute provides no remedy to the original submitter of the information and only authorises criminal prosecution of those who communicate or receive such material. A common law remedy does exist, however, for the original submitter to obtain damages for his losses.

## NORWAY

The concept of confidentiality appears in Section 13 of the Public Administration Act which requires administrative officials to protect the secrecy of official information concerning "technical devices, production methods, business analyses, calculations and industrial and trade secrets". This refers to information submitted by a private firm and which others could exploit for their business purposes.

Violators of Section 13 may be prosecuted under criminal law. An injured party may seek redress by bringing a suit against the discloser of the information in a civil court.

## SWEDEN

The concept of confidentiality is dealt with in the Official Secrets Act (SFS 1980:100, chapter 1). Under this act, information which is designated confidential may not be revealed - orally, by the disclosure of a document or otherwise - or used outside the activity concerned, unless provided for in the Official Secrets Act or in statutes to which this act refers.

These restrictions apply to any authority where information is designated confidential and to all employees or persons working for these authorities. According to the Penal Code, section 20, para. 3, a public servant who wrongfully discloses or uses confidential information will be suspended or removed from his office and, when deemed appropriate, imprisoned or fined.

## SWITZERLAND

Swiss law recognises the concept of the manufacturing and business secret ("secret de fabrication et d'affaire") and defines it as all specific knowledge which is not in the public domain, not easily accessible, and which the manufacturer or business has a legitimate interest in keeping secret for his exclusive use. Manufacturing and business secrets cover all information which can affect the business of an enterprise. Manufacturing secrets relate to a company's manufacturing facilities and processes and formulations which are not publicly known and which are valuable to the company. Business secrets include sources of supply, business organisation, calculations of prices, advertising and production. This type of information is considered confidential when submitted to the government and is not to be disclosed. In the case of wrongful disclosure, the submitter may sue for damages in civil court and may also bring a criminal complaint.

## UNITED KINGDOM

The concept of confidentiality per se is not defined in British law, nor is there specific legislation directly relating to commercial confidentiality. Under Section 2 of the Official Secrets Act (1911) it is a criminal offence to disclose official information unless authorised to do so. Some statutes require industry to submit information to the government for specific purposes and protect some of it from disclosure. For example, the Statistics of Trade Act (1947, amended) prohibits disclosure of the information it requires enterprises to provide to the Department of Industry.

With regard to data required by health and
safety legislation, the Control of Pollution Act
(1974) and the Health and Safety at Work Act (1974)
prohibit unauthorised disclosure of information submitted as chemical control legislation; they are
discussed in Section 2.2).

Breach of Confidence: UK common law recognises
that "breach of confidence" occurs when confidential
information has been disclosed without authorisation,
leading to financial loss or other injury to the submitter (*). This tort is derived from common law and,
by definition, is not based on legislation or codes.
It has, however, been suggested that breach of confidence become a tort based on statutory law.

> "A new tort (i.e. civil wrong) of breach of
> confidence should be created by statute. The
> breach of confidence constituting the tort
> should be a breach of statutory duty of con-
> fidence not to disclose nor use information
> acquired in confidence - except to the extent
> that such disclosure or use is authorised by
> the person to whom the duty is owed. It
> should include a duty to take reasonable care
> to ensure unauthorised disclosure or use does
> not take place..." (**).

To prevail in court, the submitter must show that

(i)     the information had the necessary
        quality of confidentiality about it;

(ii)    the information was disclosed by the
        defendant who had an obligation to
        maintain confidentiality; and

(iii)   there was an unauthorised use of that
        information.

A third party who receives confidential information
which he knows, or ought to know, reached him through
a breach of confidence, is legally bound not to use
or to disclose it further. The UK courts will grant
a restraining injunction to enforce this duty at the
request of the injured party.

---

(*)     Seager v. Copydex (1967) (Lord Denning).

(**)    U.K. Law Commission, Working Paper No. 58,
        "Breach of Confidence" (1974) (recommendation
        to UK Government).

The owner of the confidential information suing for breach of confidence has recourse to civil remedies: injunction to prevent further unauthorised disclosure and payment of lost profits and/or damages.

However, it must be noted that there has not been an actual case to date where a breach of confidence action has been brought against the government or a named civil servant. Therefore, this is hardly a common remedy in cases where data is submitted to government administrations.

UNITED STATES

Two general statutes which apply to all US government agencies define the concept of commercial confidentiality for information possessed by the government.

The Trade Secrets Act in the 1948 Criminal Code (18 USC 1905) makes it a crime for any officer or employee of the US government to publish, divulge, disclose, or make known in any manner not authorised by law, information which "concerns or relates to trade secrets, processes, operations, style of work, or apparatus, or to the identity, confidential statistical data, amount or source of any income, profits, losses, or expenditures of any person, firm, partnership, corporation, or association..." This law prohibits disclosure of trade information "to any extent not authorised by law". If no other law authorises the disclosure of specific commercial information, 18 USC 1905 makes disclosure a crime. It is not yet certain to what extent an injured party can use this law as a remedy, either to prevent disclosure or for redress in the event of wrongful disclosure. Still, the obligation rests on officials not to release this material, and any government officer or employee who wrongfully discloses data is subject to criminal prosecution.

The Freedom of Information Act (FOIA) 5 USC 552 contains exemptions from public disclosure which authorise the government to withhold trade information. Exemption (b)(4) concerns "trade secrets and commercial or financial information". These exemptions do not require the government to withhold information; they only authorise it to do so. The FOIA is discussed more fully in Section 1.2.

It may also be possible for an injured party to seek money damages in a claim against the United States for negligent wrongful disclosure of information. Such an action, which might be brought under the Federal Tort Claims Act (28 USC 2671 et seq.), would not prevent disclosure, but only allow the possibility of obtaining damages afterwards. In addition, an action might be brought under the Fifth Amendment of the United States Constitution seeking compensation for the "taking" of private property for a public purpose. However, these forms of redress are untested.

There are also provisions in statutes regulating pesticides, drugs, and chemical substances in general which require agencies to protect industrial and commercial secrets and which authorise disclosure only under limited circumstances. These statutes are discussed in Section 2.2.

In general, whether information submitted to a federal agency should be protected as a trade secret remains controversial and the issue is receiving consideration in Congress and the courts. Current policy may vary among agencies and sometimes among individual programmes within a single agency.

## 1.2: PUBLIC ACCESS TO INFORMATION IN GOVERNMENT FILES

In the preceding section the general statutes defining and protecting confidential information were discussed. Some Member countries require that all documents in their possession be considered confidential. Other countries take a different approach and regard the documents they possess as non-confidential in principle, and therefore accessible to the public. In the latter case confidentiality is defined in terms of exemptions to public access which either authorise an administrator to refuse access at his discretion or which simply prohibit the disclosure of certain documents.

Some public access laws are quite specific in delineating confidential information, and there are generally provisions for exempting certain industrially or commercially valuable information from disclosure to the public on demand. Where data on chemicals are concerned, these particular exemptions are relevant.

### AUSTRALIA

At the present time there is no general statutory provision in Australia giving the public freedom of access to administrative documents. A Freedom of Information Bill has been proposed and introduced in Parliament where it is still being considered. If enacted, this proposed bill would provide access to commercial or industrial documents held by administrative agencies, subject to certain exemptions for trade secrets and documents which would constitute a breach of confidence if disclosed. Another exemption covers documents which fall within the secrecy provision of other statutes.

### CANADA

Canada is working towards legislation that will allow its citizens access, with limitations, to information in government records. The most recent version of the proposed Access to Information Bill states that the Bill's purpose is to extend the present laws of Canada to provide a right of access to information in records under the control of a government institution in accordance with the principles that government information should be available to the public, that necessary exceptions to the right of access should be limited and specific and that decisions on the disclosure of government information should be reviewed independently of government.

Under the proposed Bill, trade secrets, confidential business information, confidential scientific research and technical information received from industry and foreign governments or organisations are expected to be exempt from disclosure.

DENMARK

The Public Access to Administrative Information Act (Act No. 28, 10th June, 1970) entitles any person to obtain official documents concerning any matter being considered by a public administration. For example, with regard to pesticides, a request would be filed with the Pesticides Board. The principal aim of the law is to provide the public with administrative documents, not to secure industrial and commercial confidentiality. The Act simply refers to these official documents as "documents" and makes no distinction between "administrative" documents and "industrial or commercial" documents. Section 2 of the Act denies access to information on technical applications or processes or modes of operation or matters of business in so far as it is of essential financial importance to the person or enterprise concerned that the request for disclosure not be granted.

The authorities decide at their discretion whether access should be granted or an exemption should apply. According to the exemption provision, each separate document must be considered before release. Decisions on commercial exemptions must be reached by balancing the public's interest and right to know and industry's interest in its property. The authorities may consult the party to be affected by the disclosure if some doubt exists about the interests at stake.

An important requirement of the Danish public access law is that the request for access must state the relevant specific case and document. This condition is generally interpreted as requiring that the requester have prior knowledge that the specific, relevant case exists. The request for documents must contain at least the information necessary to locate the case in the agency's normal filing system. An administrative agency is not under obligation, for example, to go through all the cases dealing with a particular area or registered during a certain period.

The general confidentiality and secrecy provisions of Danish law discussed in Chapter 1 do not
prevent disclosure of materials covered by the public
access law. However, special regulations on confidentiality do prevail over the public access law and
can prevent disclosure. There are no such special
regulations under the Pesticides Act or the other chemical control statutes. Therefore, any restrictions
on public access for chemical information will be
based on Section 2 of the public access law (discussed
above) which protects technical and business matters.

## FEDERAL REPUBLIC OF GERMANY

There is no specific German legislation
granting the general public access to administrative
documents. It is totally at the discretion of an administration to release documents in its possession.

However, two categories of persons do have the
right to examine certain documents held by an administration: (i) journalists may review official
material, as provided for in the Press Codes enacted
by the state legislature; and (ii) a party to an administrative action or procedure has the right to
review documents pertaining to his individual case.
If the administration refuses to provide documents to
a requester from either of these two categories, the
requester may file suit in an administrative court.

Once the administration grants permission to
examine the documents, it still cannot release confidential commercial and industrial information
without permission of the submitter of the data.

## FINLAND

The Publicity of Official Documents Act (83/51)
grants Finnish citizens the right of access to official documents designated as not-restricted. Only
citizens of Finland can demand such documents. Other
nationals do not have that right, but may receive such
documents at the discretion of the administration.

The Decree concerning Exceptions to the
Publicity of Public Documents (650/51), enacted in
1951, lists the types of documents that cannot be made
public. The disclosure of documents containing information or reports obtained by the administration
about private enterprise, commercial or industrial
activity, economic or professional practice or an individual's economic position requires the consent of
the submitter.

However, not all documents meeting these criteria are necessarily exempt from disclosure. Within its own discretion the agency may determine whether it is necessary to protect the submitter's interests by keeping the data secret.

## FRANCE

Law No. 78-753 of 17th July, 1978 gives the public the right to examine certain administrative documents. Information pertaining to a particular physical person ("documents de caractère nominatif") is not available to the public. The law enumerates the following categories of administrative documents: dossiers, reports, studies, records, minutes, statistics, directives, instructions, circulars, notes and answers of ministers containing interpretations of law or descriptions of administrative procedures, appraisals (excepting appraisals of the Council of State and administrative courts), forecasts and decisions.

The documents may be in writing, on sound or visual tracks, or in automatic data processing storage.

Article 6 of the law lists a number of exceptions whereby authorities may deny access to documents. One of these concerns documents containing commercial and industrial secrets ("secret en matière commerciale et industrielle", not defined further). Moreover, in fulfillment of the law, French Ministries have promulgated lists of categories of documents or data which are exempted from public access.

Article 5 provides for the creation of a commission for access to administrative documents (commission d'accès aux documents administratifs) which is empowered to adjudicate cases involving an administration's refusal to release requested documents (see also Decret No. 78-1136 of 6th December, 1978) This commission also gives advice to the Ministries promulgating lists of categories of restricted information. Article 9 of the the law prescribes the publication of certain documents at regular intervals.

## ITALY

There is no provision in Italian law granting access to administrative documents in general, but in particular cases provided for in special regulations the public has a limited right of access to preparatory acts (e.g. investigation reports) and administratative acts (e.g. regulations, authorisations).

NETHERLANDS

The <u>Government Information (Public Access) Act</u>
(1980) gives citizens the right to obtain documents
held by a government agency. In principle, this act
applies to all types of government information, including industrial and commercial documents. All
requests for government documents must be granted
unless covered by a specific exemption prohibiting
disclosure. One such exemption is for business or
manufacturing data given in confidence to the
government by natural or legal persons.

As for information not explicitly recognised as
such, the administration decides, judging from the
data themselves, whether they are sufficiently confidential to be withheld. The Act states that third
parties shall not receive any unfair advantage over
the submitter of the information as a result of disclosure.

A denial of a request for documents under this
act may be appealed to the Judicial Section of the
Council of State, as authorised by the <u>Administration
of Justice (Administrative Decisions) Act</u> ("AROB"
Act). The Act's secrecy provision does not apply to
all legislation. A number of environmental protection
statutes are governed by the secrecy provisions of the
<u>Protection of the Environment (General Provisions) Act</u>
<u>(1979)</u>.

NORWAY

The <u>Public Access Act</u> (19th June, 1970) permits
the public to obtain access to official records and
documents. Documents, for purposes of this act are
materials either prepared by an administration or submitted to it and include documents of industrial and
commercial nature. Section 6, No. 4 of the Act lists
exemptions from disclosure, such as documents which
contain information about technical devices and procedures as well as production or business matters.
This provision only grants the administrator the right
to deny access to the document.

However, Section 5, No. 3 of the same act prohibits public access to documents designated as confidential by, or pursuant to, statutory law. Various
types of trade secrets are protected by the <u>Public
Administration Act</u>, the <u>Product Control Act</u>, the
<u>Pesticides Act</u>, and more generally, the <u>Marketing
Control Act</u>. Since this material is already
designated secret by these statutes, an administration

is obligated to refuse requests for access.  However, all of the other exemptions in the Public Access Act are discretionary, not mandatory, and decisions can be appealed by requesters for information.

## SWEDEN

Chapter II of the Freedom of the Press Act (SFS 1949: 105) grants Swedish citizens access to official documents.  As used in that law, the term "official documents" refers to all documents in the hands of national or local officials, whether the documents are submitted to the authorities or prepared by them.  A document is considered prepared by the authority when it is finalised and filed.  There is a limitation to the right of free access, however, in the form of exceptions.  All exceptions to public access must be stated precisely either in the Official Secrets Act (SFS 1980: 100) or in other statutes.  One exception protects documents affecting the legitimate economic interests of collective groups and private persons.

## SWITZERLAND

Because there is no statute giving the public access to records and because of the professional secrecy provisions for civil servants, there is little access to official records in Switzerland.

## UNITED KINGDOM

There is no particular statute in the United Kingdom granting the public access to administrative documents in general.  Under the Public Records Acts (1958 and 1967) administrative documents become available, if at all, only with government approval and after a fixed number of years.  In order to be released, a document must first be designated a "public document" and subject to safeguards in particular cases; it becomes open to the public 30 years after its creation.

Certain specific statutes do contain provisions providing the public with access to particular administrative documents.  An example, relevant to chemicals control is contained in Sections 41 and 42 of the Control of Pollution Act (1974) which gives the public the right of access to registers maintained by water authorities and to the details of effluent discharges and water samples.

In addition, some industrial or commercial information may be disclosed under "Town and Country
Planning" legislation (e.g. the Town and Country
Planning General Development Order (1977). Applicants
for planning permission are required to submit to the
planning authority considerable data about their proposals. For example, in the case of a proposed
chemical plant, the authority may require detailed information about the site and design, which often involves submitting confidential industrial
information. A public local enquiry is also
authorised at which all interested persons may inspect
any of these data which are in evidence.

UNITED STATES

The Freedom of Information Act (FOIA) 5 USC 552
provides that any person may request, in writing,
records from any agency of the US Government. The
agency must provide those records unless they are
exempt under one of nine narrowly-drawn exemptions.

The fourth exemption to the FOIA, 5 USC
552(b)(4), is the only general legal mechanism for
protecting industrial and commercial secrets contained
in US administrative records. This exemption allows
an agency to avoid disclosing "trade secrets and
commercial and financial information obtained from a
person and privileged and confidential".

There is case law concerning 5 USC 552(b)(4)
which has established a general standard for commercial confidentiality. If it can be shown that the
information in question has been kept confidential by
the submitter and is not available to others through
lawful, reasonable means, the information must then
meet one of two tests: (i) either disclosure of the
information would be likely to impair the government's
ability to obtain necessary information in the future,
or (ii) disclosure of the information would be likely
to cause substantial harm to the competitive position
of the submitter. If either of these tests is met,
the information qualifies as confidential under 5 USC
552(b)(4).

A person requesting public access has a right
to an appeal within the agency if the agency fails to
provide the requested records. If the agency denies
the request for records, the person may file suit in a
United States District Court to seek disclosure. The
Court is to review de novo whether the requested
records are exempt from disclosure. If they are not
exempt, the Court will order disclosure.

Just as the party requesting the material can sue the government in the federal courts in order to compel disclosure, the submitter of the information has a legal remedy. While there would be no legal redress for an injured party in the event of wrongful disclosure under the Freedom of Information Act, if the supplier has notice of a proposed disclosure he can seek judicial review of the agency decision and obtain a court order preventing disclosure.

An unresolved question is the possible conflict between the FOIA and the Trade Secrets Act, 18 USC 1905, the other general statute protecting confidentiality (discussed in Section 11). 18 USC 1905 protects material from disclosure "not authorised by law", while the FOIA authorises, but does not compel, the government to withhold confidential data. While the question has not yet been resolved, a number of legal experts believe that 18 USC 1905 requires secrecy for information covered by both the FOIA's trade secrets exemption and 18 USC 1905.

# CHAPTER 2

## DATA ON CHEMICALS

### 2.1: NATIONAL REQUIREMENTS FOR SUBMITTING DATA AND FOR NOTIFICATION AND AUTHORISATION OF CHEMICALS

Until some fifteen years ago the chemicals control legislation of most OECD countries aimed at the regulation of specific chemicals or categories of chemicals for reasons concerning workers' protection or a recognised potential for harm to man or the environment related to the use pattern. Systems of registration and authorisation became widely used for products such as pharmaceuticals and pesticides. Applicants for an authorisation had to submit detailed information of all kinds.

More recently, and after the adoption in Switzerland in 1969 of the Federal Law on Trade in Toxic Substances, many OECD countries have opted for the anticipatory control of chemicals in a broader sense. The focus is on anticipating the effects of a chemical on man and the environment through an assessment prior to the introduction of the chemical on the market, on the basis of available information or on mandatorily required data. Although there is consistency in intent of the new generation of legislation in OECD countries, significant differences in requirements exist. In addition, the responsibility for the assessment may be shared differently between the controlling administration and the manufacturers or importers of chemicals. In many European countries, however, the implementation of EC Council Directive 79/831/EEC will lead to a great similarity in national requirements for data submission.

Of the data submitted in fulfillment of control legislation, some may be business secrets, some may require confidential handling because of proprietary rights involved and some may in fact be considered to be in the public domain. This section merely attempts to give a quick review of existing national provisions for submission of data. For further details the reader may wish to consult the various legal texts which are mentioned.

## AUSTRALIA

In general, the following classes of chemicals are subject to regulation in Australia: therapeutic substances, food additives, poisons, pesticides, veterinary drugs, agricultural chemicals, feed additives and explosives.

Under the Australian Constitution, regulation of chemicals (other than imports) lies primarily within the authority of the State administrations. However, the Commonwealth (federal government) and the States co-operate through consultative bodies such as the National Health and Medical Research Council (NH&MRC), the Australian Agricultural Council (AAC), and the Australian Environment Council (AEC).

Data submitted to the Commonwealth: the national government is authorised under a few statutes to obtain data on chemicals from industry. The Commonwealth Department of Health may require information about imported therapeutic substances under the Imported Therapeutic Goods Act, about substances falling under the Narcotics Drugs Act, and about some specific substances controlled under the Customs (Prohibited Imports) Regulations. For example, the import of polychlorinated biphenyls and terphenyls is controlled under the latter.

Information submitted by industry can be withheld by the Commonwealth under certain regulations carried out by the States. The States and the Territories rely on the Commonwealth's National Health and Medical Research Council (NH&MRC) to evaluate data submitted by those wishing to market or use food additives, pesticides and poisons.

By agreement with the States, which have the ultimate legal authority for regulatory control, the Commonwealth is enabled to co-ordinate and participate in a joint Commonwealth-State scheme for the clearance of new agricultural chemicals and veterinary drugs before they are considered by State authorities for registration. Industry is obliged to submit data to Secretariats maintained within the Commonwealth Department of Primary Industry.

Data submitted to the States and Territories: regulatory procedures within the States and Territories vary considerably. Control of chemical substances is usually exercised through Acts and Regulations relating to broad categories of chemicals. These broad categories may include poisons, explosives, commercial gases, flammable

liquids, oxidising substances, radioactive preparations, pesticides, veterinary medicines and animal
feedstuffs, agricultural chemicals and food and
drugs. Legislation does not exist in the States to
control specific chemicals. Further control may be
exercised through the Health Acts. Such Acts may be
administered by various authorities such as health or
primary industry departments.

CANADA

Canada has legislation that deals with the collection of data on industrial and commercial chemicals
(the Environmental Contaminants Act, Transportation of
Dangerous Goods Act and the Statistics Act) and other
legislation that controls chemicals in view of their
specific uses and requires registration or approval
(Food and Drug Act, Pest Control Products Act,
Hazardous Products Act).

The Environmental Contaminants Act (1976), has
a provision for the mandatory reporting of "new"
chemicals, i.e. chemicals imported or manufactured in
excess of 500 kg within a calendar year for the first
time by a company. Data to be submitted in the notification include the name of the chemical, the es-
timated quantity and any available information on danger to human health or the environment. Since this
information is not sufficient for a preliminary
assessment of a new chemical, a set of not-mandatory
Reporting Guidelines (1978) were published listing the
types of information that ought to accompany the
notification, such as structure, physical chemical
properties, biodegradation/accumulation data, amounts
and uses, toxicity data, supplier, packaging.

Other sections of the Act authorise the systematic investigation of "existing" substances or
classes of substances in order to determine their fate
in commerce and in the environment. Industry may be
required to submit data on imports and production as
well as details of processes, impurities and losses to
the environment. Furthermore, when evidence is sufficiently strong, and to improve the understanding of
the threats selected chemicals may pose, industries
may be required to carry out tests with respect to the
physical, chemical and biological properties.

DENMARK

The Chemical Substances and Products Act
(October 1980), replacing The Poisons and Substances
Injurious to Health Act and The Pesticides Act, con-
trols chemical substances and products in general.
The Act sets out an authorisation procedure for

pesticides and a notification procedure for new substances in accordance with the EC Council's Directive
79/81 EEC. Substances produced in significantly increased amounts or used for significantly altered purposes must also be notified. Notifications and
reports are sent to the Working Environment Institute
for technical registration of the data.

## EUROPEAN COMMUNITIES

All EEC member states will have to amend their
existing legislation or enact new laws to comply with
the EC Council Directive 79/831/EEC.

Of the countries discussed in this report, the
following are members of the European Communities:
Belgium, Denmark, the Federal Republic of Germany,
France, Italy, the Netherlands, and the United
Kingdom. Some of the EC member states have not yet
implemented the 1979 Directive.

The Directive requires notification of new substances by all manufacturers and importers 45 days
before they are placed on the market. A dossier must
be addressed to the responsible authority of the
country where the substance is produced or imported
into the EEC. According to Article 6, the submission
must include a technical dossier with the information
necessary to evaluate risks and test results as set
out in the Directive's Annex VII.

Second notifiers of chemical substances which
have already been notified may, with the permission of
the competent authority incorporate by reference the
results contained in the previous notification. This
requires the permission of the first notifier who
originally submitted the results. Producers must also
provide the authorities with information on new uses
and effects of a previously notified substance, new
quantities placed on the market, and any change in the
properties of the substance as a result of a mod-
ification. When the quantity put on the market by a
notifier reaches a level of 10 tons per year or a
total of 50 tons, the notifier may be required to provide the results of further studies set out in the
Annex VIII of the Directive.

A "Level 2" series of tests exists for new substances which are marketed in quantities of 1000 tons
per year or 5000 tons in total. In addition, the
authority in a Member state may request further tests
or verifications to evaluate the hazard posed by a
substance.

FEDERAL REPUBLIC OF GERMANY

The Act on Protection Against Dangerous Substances (Chemikaliengesetz-Chem G) was adopted on 16th September, 1980 (Bundesgesetzblatt I, p. 1718) in accordance with the requirement of the EC Directive 79/831/EEC.

Manufacturers and importers of new chemical substances must notify the administration of the chemical's identifying features, details on use, harmful effects during use, proposed quantity to be introduced, and methods to dispose of, recycle, and neutralise the substance. Results of tests which must also be submitted by the notifier include data on various toxic effects and determinations of physical and chemical properties. The authority responsible for receiving the notification may request further testing information for substances which are probably dangerous and for substances marketed in the EC in large quantities. The authority may also request the submission of test data for an existing dangerous chemical for which notification is not required by the EC Council Directive.

FINLAND

At the moment, chemicals are controlled under the Poisonous Substances Act plus several acts and regulations applying to certain groups of chemicals.

Under the Poisonous Substances Act, the authorities are entitled to all the information needed for supervision. Under separate acts, authorisation is needed for many categories of chemicals. A kind of notification system exists under the Occupational Safety Act which requires that a material safety data sheet on all substances and preparations hazardous to health be filled in and sent to the National Board of Labour Protection.

FRANCE

Act No. 77-771 of 12th July, 1977 on the Control of Chemicals requires premarketing not-ification of new substances and of substances presenting a new hazard due to of increased volume, modifications of the manufacturing process, or new use for the chemical. Decree No. 79-35 of 15th January, 1979 requires that the notification's technical dos-sier contain, in particular, the following in-formation: generic and trade names of the substance, chemical formula, principles of the manufacturing pro-cess, physical and chemical properties, impurities and additives, description of commercial packaging, quantities to be manufactured or marketed, known effects, types and conditions of use, results of toxicity tests and of mutagenetic pro- perties research, and degradability assessment data. The notifier may be requested to provide further test data to supplement this dossier.

The dossier is submitted to the Ministry of the Environment. Copies are passed to the Ministries of Labour and Health, and a résumé is given to the Ministry of Transport. The dossier differs very slightly from that prescribed by the EC Council Directive and will be modified accordingly in the near future.

Notification of new substances and new preparations are also required under Article L 231-7 of the Code du Travail and Decree No. 79-230 of 20th March, 1979.

## JAPAN

Under the Law concerning the Examination and Regulation on Manufacture, etc. of Chemical Substances (1973), a notification procedure has been introduced. The notification dossier must contain the following items: the name of the new chemical substance, its formula, properties, composition, use, volume of production or import, and location of the manufacturer. If appropriate, the notifier adds test reports on biodegradability, bioaccumulation or long-term toxicity of the notified chemical. Authorisation is required for the manufacture or import of chemicals with persistant and bioaccumulative and toxic properties. This requirement applies to both existing and new chemicals.

## THE NETHERLANDS

The Chemical Substances Bill, drafted in view of implementing the notification procedure of EC Council Directive 79/831/EEC, was presented to Parliament for consultation on 15th May, 1981. The Bill provides for mandatory notification by anyone intending to manufacture or import a new chemical into the Netherlands or to put a new chemical into circulation. The Minister of Health and Environmental Protection is responsible for receiving the notifications. Copies of the dossiers and further data received are to be passed to the Minister of Social Affairs, as well as to the Minister of Transport and Public Works, if they are of importance with respect to the transport of the chemical.

## NEW ZEALAND

The Toxic Substances Act (1979) lays down compulsory authorisation or notification procedures for
toxic substances. It authorises the Director of
Public Health to request details of composition, purpose and method of use for any toxic substance.
Licenses are required for the sale and distribution of
poisons and poisonous substances.

Among other statutes in this area is the
Pesticides Act (1979) which requires that applications
for registering pesticides be accompanied by information concerning composition, analysis, and efficacy data and any other data specified by the
authorities. It also requires information on the
analysis of residues and on the effects on wildlife.

## NORWAY

The Product Control Act (1976) gives the
authorities general responsibility for all chemicals,
products and substances which may constitute a danger
to man and the environment. Under the Act, the King
may rule that products may not be produced, imported
or sold without prior authorisation. The principle to
date has been that manufacturers, importers and
dealers must obtain authorisation before producing,
importing, exporting, or selling poisons and
healththreatening chemicals.

Compulsory authorisation or notification procedures for chemicals in general have not yet been established under the Product Control Act, but this may
be done by Royal Decree. The Act also authorises the
administration to demand information regarding a product and to make inquiries about the effects and qua-
lities of chemicals.

In accordance with the Act Relating to Worker
Protection and Working Environment and the Product
Control Act, a registration system for chemicals
manufactured in or imported in to Norway is to be established. This registration system relates to
chemicals which are classified as toxic, hazardous to
health or which have special long term effects. This
system is organised by the Product Register.

SWEDEN

The manufacture of and trade in hazardous chemical substances and products are covered by The Act on Products Hazardous to Health and to the Environment (SFS 1973:334). Under the Act, a manufacturer or importer has the obligation to be fully knowledgeable about a chemical's composition and its effects on health and the environment. According to the preparatory documents to the Act, exceptions are allowed only when it is obvious that the product does not present any risk to health or the environment. Before the chemical enters the market, it must be clearly marked with all the information relevant to public health or environmental protection.

Like most countries, Sweden has a registration system for manufacturing, marketing and use of pesticides. Commercial manufacture of and trade in poisons (substances carrying grave health risks in use) require a permit issued by the County Administration. Anyone intending to use poisons in his occupation must notify the Industrial Safety Inspectorate. The commercial manufacture of and trade in substances which represent health risks of a lesser degree than poisons should also be notified to the Industrial Safety Inspectorate.

Otherwise, Sweden has no formal notification system for new chemicals. The Products Control Board may, however, require all information needed to assess any chemical, such as toxicological and environmental data and information about a chemical's components, including impurities. The Board keeps a register of chemicals manufactured in and imported into Sweden. The supervising authorities, the National Environment Protection Board and the National Board of Occupational Safety and Health, may request all the information needed to supervise compliance with the law.

SWITZERLAND

Toxic substances are regulated under the Federal Law on Trade in Toxic Substances ("Loi fédérale sur le Commerce des Toxiques") (1969) which contains provisions for authorisation and registration. The Swiss chemical law draws a distinction between trade and marketing in toxic substances. Trade is the manufacturing, pro- cessing, storing, use, import, sale, purchase, re- commendation, or offering for sale or disposal. Marketing, for purposes of this law, is the initial manufacture or import and initial recommendation or offer for sale in Switzerland.

With a few exceptions, a permit is required to trade in toxic substances. The cantonal authorities, who are generally responsible for supervising toxic substances within their jurisdiction, issue the authorisation permits. Toxic substances approved for trade are entered into an official register (the list of toxic substances). Persons trading in toxic substances must provide all data needed to classify a chemical upon request. This includes permitting the authorities to review pertinent documents and conduct on-site inspections.

A toxic substance may not be marketed until it has been included in the list of toxic substances. Persons intending to market a toxic substance not yet appearing on the list must file a declaration with the Federal Office of Health. The declaration must contain information on the substance's composition, intended purpose or application, method of use, and any scientific test and investigation results or advertising material (Implementing Order 23rd December, 1971, Article 18).

## UNITED KINGDOM

The United Kingdom has compulsory authorisation and notification procedures for certain chemicals. Under the Health and Safety at Work (etc) Act (HASAWA) (1974), regulations may be promulgated requiring manufacturers or importers of chemical products to notify the authorities about certain properties of substances they intend to manufacture or import. Regulations will be issued shortly to implement within the UK the EC Directive 79/831 EEC.

The United Kingdom also has compulsory authorisation procedures. For example, under the amended Alkali (etc) Works Regulation Act (1906), production of certain chemicals may not be carried out unless the works in which the process takes place are registered. The producer must also obtain an annual certificate from H.M. Alkali and Clean Air Inspectorate.

## UNITED STATES

The Environmental Protection Agency may require industry to submit various information under sections of the Toxic Substances Control Act (TSCA) 15 USC 2601 et seq, the Federal Insecticide Fungicide and Rodenticide Act (FIFRA) 7 USC 136 et seq. (as amended).

(i)    Toxic Substances Control Act

        Section 4 of TSCA (15 USC 2603) directs EPA to
require, through its rulemaking procedure, testing of
chemical substances if insufficient data exist to determine whether the substance poses an unreasonable
risk of injury to human health or the environment.
The data required should lead to a determination of
whether the substance presents such an unreasonable
risk of injury.  Such data may include data on car-
cinogenesis, mutagenesis, teratogenesis, behavourial
disorders, cumulative or synergistic effects, and
other factors.

        The manufacturer or processor of the substance
is responsible for conducting the tests.  If the data
already exists or is being developed, an exemption may
be granted.  However, the manufacturer or processor
receiving the exemptions may be required to pay a fair
reimbursement to the original tester.

        Section 5 of TSCA requires manufacturers and
processors of chemical substances which are new (or
the use of which is significantly new) to notify EPA
at least 90 days before the manufacture, import, or
new use begins.  A notice must be submitted containing
information known to, or reasonably ascertainable by,
the submitter.  This information includes:  name and
molecular structure, categories of use, amounts to be
manufactured or imported, byproducts, exposure, method
of disposal, test data in possession of the submitter
(such as Section 4 data) and a description of any
other data related to the effects of the substance on
health or the environment.

        The  notification,  together  with  information
concerning data submitted with it, are published in
the Federal Register subject to the Section 14 disclosure restrictions (discussed in Section 2.2).  Each
month EPA publishes a list of chemicals which have
been notified.

        Under Section 8  of TSCA, EPA may require by
rule that manufacturers or processors of chemical substances maintain records for submission including the
following data:

        .  common  or trade  name,  chemical  identity,
           molecular structure;
        .  categories of use;
        .  amounts to be manufactured or processed (in
           total and by categories of use);
        .  description of byproducts;
        .  all existing data on the environmental and
           health effects;

. information on worker exposure to the substance; and
. disposal methods.

This information essentially is the same data included in the notification required by Section 5.

Section 8 also requires manufacturers, processors or distributors of substances or mixtures to maintain records of "significant adverse reactions to health or the environment" to be determined by an EPA rule. The special records include employees' adverse reactions; consumer allegations of injury, reports of occupational disease or injury, and complaints from anyone about damage to the environment. Upon request of EPA, copies of these records must be submitted.

Section 8 moreover requires manufacturers, processors or distributors to provide EPA with health and safety studies they have conducted or with which they are familiar.

(ii)    Federal Insecticide, Fungicide, and Rodenticide Act (FIFRA)

FIFRA, 7 USC 136 et seq (as amended) requires various information about pesticides registered for distribution or sale. Section 3 requires that all pesticide registration applications contain the name of the pesticide, a copy of its labelling; a statement of all claims to be made for it, any directions for its use and the complete formula of the pesticide. If requested by EPA, the applicant must also submit all test results upon which the claims for the pesticide are based.

Under Section 8, EPA is authorised to require by rule that producers maintain records concerning the pesticides they produce. However, no records may be required pertaining to "financial data, sale data other than shipment data, pricing data, personnel data, and research data" other than those relating to registered pesticides or those awaiting registration".

## 2.2: EXISTING PRACTICES FOR CONFIDENTIAL HANDLING OF DATA ON CHEMICALS

In addition to the general provisions in the laws of Member Countries which apply to all information held by a national administration, other provisions exist which pertain to the confidentiality of data submitted under chemical control laws. These are reviewed in this section. Of particular interest are the provisions for the treatment of confidential data in countries where procedures of notification of new chemicals exist. In the few instances where the chemical laws themselves contain no provision for confidentiality, but where a definite policy on handling information on chemicals exists (e.g. in Australia and Japan), that policy has been noted. Using the material available in the responses to the questionnaire an attempt has been made to identify answers to questions such as:

- which data are regarded confidential?

- who makes the decision on confidentiality?

- what are the provisions for public disclosure of data?

### AUSTRALIA

As mentioned in Section 1.1, all data for chemicals regulation received from industry by the Commonwealth are regarded as confidential. Even the fact that a submission has been made by a firm is in itself confidential information. Security measures are taken to ensure the secrecy of data at all stages of processing, review, and storage, and at no point in the procedure are data available to the public. Commonwealth officials are obligated by the Crimes Act and Public Service Act to respect the secrecy of information in this category.

At the State level, where most of the chemical regulation is carried out, the practice of protecting confidentiality varies from State to State. Provision is made in the legislation of some States to protect the confidentiality of certain trade secrets or data. In general, some means exist to protect the data submitted by firms and in some States there are procedures for appealing decisions made on confidentiality. An administrative body reviews the question in the first instance, and further appeal can be made to a stipendiary magistrate.

In general, there is no compulsory publication of data, with the exception in certain cases of the name and description of a product which may not be sold. Under poisons legislation, certain details of toxic constituents may be required to be displayed on labels, according to the Poisons Schedule. In the case of agricultural chemicals and veterinary drugs, the only information which must be published are the results of assessment, e.g. the Poison Schedule (or classification) which is assigned to the product on the basis of toxicological data submitted and the maximum residue limit allowed on crops intended for human consumption. It is felt that publication of this information does not place manufacturers at a commercial disadvantage.

## CANADA

Some of the pieces of legislation referred to in Section 2.1 have addressed the issue of confidentiality of data directly (Environmental Contaminants Act, Transportation of Dangerous Goods Act, Statistics Act, Hazardous Products Act), while others such as the Pest Control Products Act and Food and Drug Act have not. The Departments administering the latter two Acts have either published policy statements on handling and disclosure of confidential information or have internal departmental procedures in effect. For both Acts disclosure of data is tightly controlled and no information may be disclosed without prior consultation with the submitter. The ownership of data submitted to government is a complex legal matter.

Under the Statistics Act, information that is identifiable with an individual person, business or organisation cannot be disclosed. Therefore, the policy is to disclose information such as quantities of chemicals exported, imported or manufactured only if three or more companies are involved - more than three if one company has a large percentage of the market. Disclosure is allowed if the company supplying the data gives written permission.

Both the Environmental Contaminants Act and the Transportation of Dangerous Goods Act, which require submission of industrial and commercial data on chemicals, have provisions for the protection of confidential information.

Subsection 4(4) of the Environmental Contaminants Act states that:

"Any information that relates to a formula or process by which anything is manufactured or processed, whether patented or not, or to other trade secrets or that is sales or production information that has been specified, in writing, as information that is given in confidence shall not be disclosed except as may be necessary for the purpose of this Act."

However, data reported pursuant to the "new" chemicals notification requirement are not included under the non-disclosure clause. For this reason the Reporting Guidelines referred to in Section 21 include a clause guaranteeing the same assurance as that provided by subsection 4(4) of the Act.

In practice, therefore, whether for new or existing chemicals, industry indicates initially the data for which it claims confidential handling; these claims need not necessarily be followed by government, and a court may ultimately be required to provide a final decision.

Under the Environmental Contaminants Act there is no mandatory publication of data obtained during systematic investigations or pursuant or "new" chemical notification. If public disclosure or confidential information is necessary for purposes not covered by the Act, the administration consults with the submitter of the data.

DENMARK

There are no special regulations for protecting the confidentiality of chemical data in particular. Neither the Pesticides Act nor the Chemical Substances and Products Act has any special provisions for trade secrets or data. Part 12, Section 58 of the Chemical Substances and Products Act specifically makes the officials who enforce it subject to the professional secrecy provisions of the Civil Penal Code (Section 152) discussed in Section 1.1.

EUROPEAN COMMUNITIES

The EC Council Directive (79/831/EEC) contains provisions on confidentiality, the major points of which are as follows:

(i)     Any information concerning commercial exploitation or manufacturing received by an EEC Member State and the Commission must be kept secret (Article 7).

(ii)    The notifier of the data may indicate
        the information which he considers "com-
        mercially sensitive" and which, if dis-
        closed, "might harm him industrially and
        commercially". He must provide full
        justification for his claim for con-
        fidentiality. The administration
        receiving the justification will then
        decide which information should be con-
        sidered confidential (Article 11).

(iii)   secrecy shall not apply to:

        .   the trade name of the substance,
        .   certain physico-chemical data
        .   possible ways of rendering a sub-
            stance harmless,
        .   the interpretation of toxicological
            and ecotoxilogical tests, and the
            name of the body responsible for the
            tests,
        .   recommended methods and precautions
            for handling, storage, and transport,
        .   emergency measures in case of
            poisoning or accidental spillage
            (Article 11).

(iv)    The national administration which
        receives the notification may request
        the Commission to include a substance's
        name in encoded form on the list of all
        notified substances provided that the
        substance is not dangerous within the
        meaning of the 6th Amendment. A sub-
        stance can be kept on the list in en-
        coded form no longer than three years
        (Article 11).

(v)     All confidential information in the
        hands of the Commission or a Member
        State must be kept secret. However,
        this information can be given to the
        designated notification authorities in
        other EEC Member countries, as well as
        to persons directly involved in ad-
        ministrative or legal proceeding to con-
        trol substances placed on the market
        (Article 11).

The EC Council Directive's provisions for controlling confidentiality of data in dossiers sent to
other countries is discussed in Section 3.2.

## FEDERAL REPUBLIC OF GERMANY

Under Article 12 of the Act on Protection against Dangerous Substances, the administration, at the request of the notifier, will designate "manufacturing or trade secrets" as confidential information. The following data, however, cannot be considered confidential: the trade name of the substance, the substance's physico-chemical properties, disposal, recycling and neutralisation methods, the notifier's mandatory recommendations for precautionary and emergency handling and for packaging and labelling, the evaluation of the toxicological and ecotoxicological tests and the name of the person responsible for the tests.

## FINLAND

Section 12 of the Poisonous Substances Act prohibits any official or inspector from supplying to outsiders or from exploiting for his own private benefit any commercial or professional secrets belonging to a business entrepeneur which come to his knowledge in the course of his duties. Most of Finland's statutes pertaining to specific chemical products also have some provision to protect trade secrets.

The Pharmaceuticals Act (374/35), the Foodstuffs Act (526/41) (under which, e.g., cosmetics and detergents are regulated), the Act on Substances Involving the Danger of Explosion (263/53) and the Pesticides Act (327/69) all contain a secrecy clause identical to that in the Poisonous Substances Act. Similar provisions are also contained in other regulations under which information on chemicals is either submitted to or otherwise obtained by authorities.

## FRANCE

There are provisions under the French Act 77-771 on the Control of Chemicals for protecting the confidentiality of some of the data required for notification. Protected data include information on manufacturing processes as well as projections for production and marketing. A notice published by the environment administration indicates that the chemical identity of a substance may also be protected if justification is given. However, the chemical identity of a substance cannot be considered confidential if it appears on the list of substances harmful to man or the environment. Although the enterprise makes the initial choice of the data it considers confidential the final decision is made by the administration. A submitter may appeal decisions on confidentiality to the administrative courts.

The Ecotoxicity Evaluation Commission which was created by Article 9 of the Decree 79-35 of 15th January, 1979, examines the notification dossiers, maintains the confidentiality of data it receives. Persons having access to the files are bound by the duty of administrative secrecy, except in court proceedings.

The Chemicals Control Act 77-771 requires that toxicological data contained in the dossiers be published in an appropriate form. Such publication is undertaken on the advice of the Ecotoxicity Commission to prevent erroneous interpretations or the unfair use of the data by other firms. Publication only takes place after the substance is actually put on the market. Except for substances whose recognised danger requires official classification by generic name, substances are identified at the submitter's request by trade or code names.

The Decree 79-35 of 15th January, 1979 implementing the chemical law refers to a special procedure available to industry for transmitting manufacturing secrets ("secrets de fabrication"). Normally, industry must submit five copies of the notification dossier to the Ministry of Environment for distribution among other relevant ministries. Under a special procedure, a single copy of the confidential data may be submitted separately from the rest of the dossier to the Product Control Division of the Ministry of the Environment. The data will not be circulated beyond the Ministry unless requested specifically by the other parts of the administration. The original submitter of the data cannot interfere with this transmission.

The authorities may transmit to poison control centres the full formula of chemical substances registered by producers. However, pursuant to Article 7 of the Decree of 77-1558 of 18th December, 1977 governing cosmetics and personal care products, these formula are considered secret and are protected from disclosure.

JAPAN

There is no specific provision regarding confidential data in the Japanese chemical law. As mentioned in Section 1.1, public officials are obliged not to divulge secrets acquired in their profession.

As a rule, data submitted by industry to the government in fulfillment of provisions of the Chemical Substances Control Law are not disclosed. However, the names of new chemicals which have been approved for manufacture or import are made public.

## THE NETHERLANDS

The Chemical Substances Bill, presented on 15th May, 1981, for consultation by Parliament, provides for the possibility of public inspection of parts of the notification dossiers on new chemicals for which an evaluation of the possible effects on man and the environment is considered necessary and which are excluded from secrecy in accordance with article 11 of EC Council Directive 79/831/EEC. During a specified period anyone may make written observations relating to the documents displayed.

The provisions for protecting the confidentiality of data in the Bill overrule the general provisions prohibiting disclosure of business or manufacturing data given in confidence to the government in the Government Information (Public Access) Act, as described in Section 1.2. The provisions for confidentiality of data submitted under this Bill are in accordance with article 11 of the EC Council Directive 79/831/EEC. Notifiers claiming confidentiality for data submitted are requested to present a separate text containing the data they consider non-confidential. This text, however, must offer sufficient information on the effects of the chemical on man or environment, and approval of the Minister is required. The Minister makes the decision on any claim for secrecy within a period of one month from the date of receipt. If a claim for secrecy is completely or partially refused, the Minister suspends publication of this information until the decision on the claim has become irrevocable. A denial of a claim for secrecy may be appealed to the Judicial Section of the Council of State, authorised by the Administration of Justice Act.

The Bill contains the provision that any person concerned in its implementation, having at his disposal information which he knows or may reasonably assume to be confidential, must keep such information secret unless other regulations apply or the need to publish such data arises in connection with implementation of the Bill.

## NEW ZEALAND

In New Zealand, the notifying firms determine the data to be treated as confidential. Any information submitted by an enterprise under the notification procedure will be treated by the Director of Public Health as confidential at the submitter's request. The data can be disclosed by officials for purposes of administering the Toxic Substances Act as well as for investigating or prosecuting any suspected offences. All requests from the general public or other groups for information on a notified chemical are referred to the original submitter of the data. Nothing is disclosed without the firm's specific permission.

The general remedy for wrongful disclosure of trade secrets is a civil suit. However, there is usually a clause in the chemical statutes exempting officials who act in good faith and with reasonable care in carrying out the statute's provisions if they disclose confidential data.

Other chemical-related statutes with confidentiality provisions include:

(1) The Pesticides Act (1979): All members of boards and committees must treat as confidential all information submitted by an applicant for registration of a pesticide. Information can be disclosed only to other members of the respective board or committee.

(2) The Animal Remedies Act (1967): Members of the Animal Remedies Board or its advisory committees may not divulge information obtained under the Act which is not generally available to the public. Information received under the Act is, however, available to the Ministry of Agriculture and Fisheries. After contacting the licensee in writing, the Board may publish the results of certain experiments or tests. With the approval of the Minister of Agriculture and Fisheries, the Board may also publish the full details of any components of animal remedies or the results of any analysis of a remedy, if it is in the public interest.

(3) Food and Drug Act (1969): While there is no specific clause concerning confidential data obtained under notification procedures, there is a provision protecting information obtained under the authority to inspect suspected offences.

NORWAY

Aside from the general confidentiality provisions of the Public Administration Act, two chemical
statutes in Norway protect confidential data.

Under Section 11 of the Product Control Act,
the following information obtained by officials must
be kept secret: the product's composition,
characteristics, production methods, research results,
future plans, prognoses, business analyses and calculations and business secrets in general.

The term "product" is defined to include
chemical substances as well as preparations. To
qualify as confidential information, the data must be
able to be exploited by other entreprises for their
own use or to the disadvantage of the submitter.
There is no conflict between this law and the Public
Administration Act since the latter covers "business
secrets" which includes the specific data listed above.

Section 12 of the Pesticides Act requires that
officials carrying out "duties pursuant to the Act"
must not disclose certain information. Information
graded confidential not only includes industrial and
business secrets but also any other information not
generally known. Unlike under the Product Control
Act, there is no requirement under this act that in
order to qualify as confidential the information could
be exploited by others, or used to the submitter's
disadvantage. However, information that would not
qualify for an exemption under the Public Access Act
cannot be witheld.

SWEDEN

The Official Secrets Act (SFS 1980:100, section
8, para. 6) and the Official Secrets Ordinance (SFS
1980:657, para. 2 and annex p.52) protect data arising
from public investigation or authorisation and
regulatory activities, such as those collected under
the Act of Products Hazardous to Health and the
Environment. Data on business or operating conditions, research or inventions by private persons,
enterprises or associations are to be kept secret,
provided that it can be assumed that the disclosure of
such data would harm the person or enterprise involved. Such data may not be released without the
consent of the submitter until 20 years after
preparation.

The administration decides which data are to be considered confidential. A request by the submitter for confidential handling of the data may, however, be important in the decision of the administration. The government may, in specific cases, order release of confidential data if considered necessary.

## SWITZERLAND

Article 29 of the Federal Law on Trade in Toxic Substances 814.80 (21st March, 1969) requires inspection and compliance officials, members of the Federal Commission on Toxics, and expert committees to maintain secrecy in- formation on chemicals. The advisory commission of independent experts consulted by the Department of the Interior during appeals of toxic registration decisions must also keep secret any information it receives.

Under Article 19(2), the Federal Council will determine the conditions under which the Public Health Service can relate data on a product's composition to anti-poison centers, which, in turn, are not permitted to release these data.

As a rule, data related to manufacturing and business secrets are treated in confidence. The sub-mitter of the data decides, in the first instance, which data must be kept confidential. The courts will decide the confidentiality question if a dispute arises between the submitter and the administration.

## UNITED KINGDOM

There is no legislation in the United Kingdom to protect confidential information specifically sub-mitted for purposes of authorisation. As noted in Section 1.1, the Official Secrets Act makes unauthorised disclosure of official material a criminal offence.

Information submitted for notification under The Health and Safety at Work Act (HASAWA) (1974) is protected from disclosure. The Health and Safety Executive and The Health and Safety Commission may not disclose any information without the consent of the submitter. Officials may, however, submit information to other government departments or authorities (see Section 3.1). Persons who receive such information may only use it only for official purposes. However, the authorities may order disclosure of information necessary to prevent injury or danger to workers.

## UNITED STATES

The statutes administered by EPA in the area of chemicals control are the Toxic Substances Control Act (TSCA) 15 USC 2601 et seq., the Federal Insecticide Fungicide and Rodenticide Act (FIFRA) 7 USC 136 et seq., and the Federal Food, Drug and Cosmetic Act, 21 USC 301 et seq. (partially administered by EPA). These laws have their own provisions for the confidentiality of submitted data. However, in general, they only give limited authority to disclose information to individuals, corporations and the general public.

Under EPA's confidentiality regulations (40 CFR Part 2) non-confidential data are routinely disclosed to anyone. Confidential information can be disclosed to other agencies of the US government in connection with their official duties [40 CFR 2.209 and 2.306(h)]. Confidential information may also be disclosed to contractors performing work in connection with these laws, but they must agree to keep such information secret. For these types of disclosure, EPA does not have to obtain the consent of the firm which submitted the data. The submitter, who is given notice before disclosure, can seek judicial review during the notice period of the decision to make the disclosure.

The laws usually provide a criminal penalty for any US government officer, employee, contractor or contractor's employee of the who wrongfully discloses confidential information [Section 14(d) of TSCA; Section 10(f) of FIFRA]. This is an action by the Government against the person who has committed the crime. In addition, EPA includes specific provisions in contracts which authorise access to confidential information, under which the submitter has a right to bring a court action against the contractor for violation of the contractor's confidentiality obligations. There is no other recourse against those who might benefit from disclosure.

(i)     The Toxic Substances Control Act (TSCA):

Section 14(a) of TSCA prohibits disclosure of information that falls within the exceptions to public access contained in the Freedom of Information Act (FOIA) 5 USC 522(b)(4). While the FOIA authorises agencies to withhold certain trade secrets, TSCA specifically requires EPA to withhold this same information.

Section 14(b) does not prohibit the disclosure of any of the following: health and safety data, any data reported to or obtained by the Administrator of EPA for a health and safety study concerning a chemical offered for commercial distribution, or data on a chemical subject to Section 4 testing or Section 5 premanufacture notification requirements. However, any information which would reveal confidential processes used in manufacturing or processing a chemical or mixture or which would disclose the concentration of chemicals in a mixture cannot be released under Section 14(b).

EPA determines which information qualifies to be protected under Section 14. In the event of a dispute between the Agency and the submitter, the submitter can seek review in a United States District Court. Independently of Section 14 and in accordance with Sections 4 and 5, the Administrator must publish a notice in the Federal Register which identifies the chemical substance, lists the uses and describes the tests performed and the data submitted. The chemical in the FR notice can be identified by a generic name.

Section 14 also authorises disclosure of confidential information relevant in a proceeding under the Act. The Act states that the disclosure must be done in "such a manner as to preserve confidentiality to the extent practicable without impairing the proceeding". Section 14 also authorises disclosure when necessary to protect health or the environment against an unreasonable risk of injury. This would also be some type of public disclosure, but the regulations implementing the Act in 40 CFR Part 2 provide that the disclosure must be made "in a manner that preserves the confidentiality of the information to the extent not inconsistent with protecting health or the environment against unreasonable risk of injury" [40 CFR 2.306(h)].

(ii)     The    Federal    Insecticide,    Fungicide,    and
         Rodenticide Act (FIFRA)

Section 10(b) of the Act prohibits disclosure of information which "contains or relates to trade secrets or commercial or financial information obtained from a person and privileged or confidential". EPA interprets FIFRA Section 10(b) to prohibit disclosure of information which would be exempt from disclosure under the FOIA [5 USC 552(b)(4)].

In addition, Section 10(d) provides that "safety and efficacy" data for registered pesticides are not confidential under Section 10(b). This does not apply to information which would reveal confidential manufacturing or quality control processes, confidential details of methods for testing, detecting, or measuring the quantity of any deliberately added inert ingredient, or confidential formulation.

Section 3 of FIFRA does not require EPA to publish safety and efficacy data, but simply to make it available to the public upon request within 30 days of registration of the pesticide. EPA determines which information qualifies for protection under Section 10. In the event of a dispute between EPA and the data's submitter, the submitter can seek review in a US District Court.

Section 3 states that companies who seek to register a product already registered by someone else must pay compensation to the original submitter of the data. This reimbursement is not for the disclosure of the data, it is for the use of the data under the Act. An original submitter of data is entitled to 15 years of exclusive use and reimbursement for data submitted after January, 1970, and for 10 years for data submitted after September, 1978.

(iii) Federal Food, Drug and Cosmetic Act

Since 1970, EPA has administered the section of this Act pertaining to registration of pesticides in or on food. Section 408(f) of the Act prohibits disclosure of any data submitted by an applicant for tolerance for a pesticide in or on food until the Agency publishes a rule granting tolerance or exempting the substance from tolerance [5 USC 346a (f)]. All information contained in a tolerance petition is considered confidential.

With regard to pesticide tolerance petitions filed under this law, EPA's confidentiality rules (40 CFR Part 2) are applicable. Confidential data submitted under the Act can be disclosed to other agencies as well as to contractors working on projects related to this law or to FIFRA.

The rest of the Federal Food, Drug and Cosmetic Act 21 USC 301 et seq., is administered mainly by the Food and Drug Administration (FDA). Section 708 of the Act describes when the FDA may release commercial information considered confidential and exempt from public disclosure under the Freedom of Information Act (FDIA) [5 USC 552(b) (4)]. By inference, the FDA cannot otherwise release such information. Section 708 does not apply to all data submitted to FDA, only to data falling within 5 USC 552 (b) (4)'s trade secrets definition. The administration may release this data to government contractors working on projects related to FDA's regulatory duties, but must require that the data's recipient adheres to security measures (5 USC 379).

The FDA has only limited authority to disclose confidential information to individuals, corporations, and the general public. Non-confidential data disclosed routinely to anyone under the FDA's confidentiality regulations (31 CFR Part 20). Confidential information may be released to other agencies of the US Government which have concurrent jurisdiction and separate authority. As noted, contractors may also receive data in some instances. FDA does not require the submitter's consent for these types of disclosure.

In general, "trade secrets and privileged or confidential commercial or financial information" submitted to FDA are not available for public disclosure. "Trade secrets" are defined as "any formula, pattern, device or compilation of information which is used in one's business and which gives him an opportunity to obtain an advantage over competitors who do not know or use it" [21 CFR 2061(a) and (b)].

Commercial or financial information that is privileged or confidential means "valuable data or information which is used in one's business and is of a type customarily held in strict confidence or regarded as privileged and not disclosed to any member of the public by the person to whom it belongs" [21 CFR 2061 (c)].

Information falling within these two categories will be withheld by FDA even if otherwise available for disclosure [21 CFR 20.60(a)]. In cases where the confidentiality of certain chemical information or data is uncertain and a request has been made for its disclosure, FDA will consult the submitter before deciding the issue [21 CFR 20.45].

As for drug data, FDA will not publicly disclose any data contained in a new drug application dossier until the application has been approved. Upon approval of the application, the following data are available to the public: summaries of safety and effectiveness, defined as including all studies and tests of a drug on animals and humans, and all studies and tests on the drug for identity, stability, purity, potency and bioavailability [21 CFR. 314 14 (i)]; product experience reports; adverse reaction reports; consumer complaints and a list of all active ingredients [21 CFR. 314.14(e)]. If the application is not approved or is withdrawn, all safety and effective data not previously disclosed to the public may be released.

The following data in a new drug application are not available unless (i) they were previously disclosed to the public, or (ii) they concern abandoned products or ingredients and are no longer trade secrets or confidential commercial information:

- manufacturing methods or processes, including quality control procedures;

- product, sales, distribution and similar information unless it does not reveal confidential material;

- quantitative or semiquantitative formulae [21 CFR 314.14(g)].

# CHAPTER 3

## OFFICIAL TRANSMISSION OF CONFIDENTIAL INFORMATION

### 3.1: TRANSMISSION OF CONFIDENTIAL INFORMATION BETWEEN AGENCIES WITHIN A COUNTRY

Administrative practice of exchanging information between different parts of the administration within a country may vary from one country to another. In some OECD Member countries statutory obligations exist for one part of the national administration to transfer certain types of information to designated other parts; in other Member countries merely the authorisation to do so is given; and in many countries information can only be transmitted on request. Where confidential data are transmitted, the general practice is that the recipient authority is required to keep the data confidential.

As a consequence, the consent of the original submitter of the data is generally not required. Frequently, statutory provisions limit the use of the data by the recipient authority to the purpose for which the data were originally submitted.

This section collects some information of particular interest derived from national legislation or administrative practice.

In Australia the information contained in submissions handled at the level of the Commonwealth and relating to certain categories of chemicals is communicated to the State government departments through the existing joint Commonwealth/State assessment bodies.

In Canada, subsection 4(4) of the Environmental Contaminants Act dealing with non-disclosure of data, does not prohibit the transfer of confidential data among departments, provided it is for "purposes of the Act". In practice, copies of detailed dossiers received under the mandatory notification of "new" chemicals are sent to an Interdepartmental Evaluation Committee for review, provided that the member departments have appropriate safeguards in effect for protection of the data. Such transmittals do not require the consent of the submitter. In other situations under the Act, the national authority may feel obligated to inform the submitter prior to transmittal. Under the Statistics Act there are provisions for data exchange agreements with other departments and for joint data collection, provided these agencies are legally bound to observe the same secrecy requirements as the administering authority. In general and in respect of all legislation, it is very difficult for one government department to obtain confidential information from another department for purposes other than those for which it is collected.

A main principle in Danish administrative practice allows departments of administration to transmit information to other departments on request.

France has special protection procedures for confidential chemical information which require the Ministry of the Environment to provide data to other ministeries upon specific request. Information from the notification dossier is also sent to anti-poison centres, where it is treated in confidence (Decree 77-1558 of 28th December, 1977).

In Japan, Article 3.2 of the Law concerning the Examination and Regulation of Manufacture, etc., of Chemical Substances requires the Ministries of Health and Welfare and of International Trade and Industry to provide the Environmental Agency with the copy of the notification dossiers, whereby confidentiality is maintained.

In the Netherlands, Section 24 of the Air Pollution Act and Section 12 of of the Protection of the Environment (General Provisions) Act require officials to provide other agencies with all information relating to authorisations, including manufacturing and industrial data.

Norway's Product Control Act obligates agencies to provide other officials with information necessary to implement the Act. The obligation also applies to confidential information, regardless of secrecy regulations.

In Sweden, certain authorities are under a statutory obligation to release information to other authorities. There is, however, no provision stating the right of authorities to obtain classified information from other authorities.

In <u>Switzerland</u>, Article 77 of the <u>Decree</u> of 23rd December 1971 implementing the <u>Federal Law on Trade in Toxic Substances</u> provides that anti-poison centres and cantonal officials receive data absolutely necessary to fulfill their duties, such as confidential information on the composition of products.

The <u>United Kingdom's Health and Safety at Work Act</u> permits disclosure of confidential data when necessary. The authorisation appears in Section 28 as an exception to the rule that the submitter's permission is required to release confidential information to third parties.

The information may be released to a government department or local authority, but only for the purposes of that department or authority, or to any other person, but only in cases where the UK Health and Safety Executive finds it necessary to do so in performance of its functions to assure health and safety in the workplace.

To receive confidential information, the recipient official agency must be a government department or an official body performing a function of the UK Health and Safety Executive, the Health and Safety Commission or other Department.

In the <u>United States</u>, Section 14 of <u>TSCA</u> (as noted in Section 2.2) authorises EPA to furnish confidential information to other US government agencies to carry out health and safety environment or law enforcement functions under other statutes. EPA, however, is not obligated to provide such information on request; it is merely authorised to do so. The recipient agency is required not only to keep the information secret, but also to have adequate security to protect it.

## 3.2: INTERNATIONAL TRANSMISSION OF INFORMATION

To date, no country has granted its officials authorisation to transmit confidential information in the absence of bilateral or multilateral international agreements. Clearly, one reason for this is that, while national governments may feel secure in permitting their agencies to transfer data to other agencies under the same secrecy restrictions, they have not yet arranged a similar guarantee of protection from foreign administrations. While by no means insurmountable, there are naturally more complications and obstacles in developing a system for protecting confidential data once the data leave a government's possession.

In this regard, the data circulation system designed by the EC in its EC Council Directive (79/831/EEC) is an interesting model. Under the EC Directive, a Member state is required to send the Commission a copy of the notification dosier, or a summary of it, along with the results of any further supplementary testing required (Article 9). The Commission then forwards to the other EC Member states the dossier (or summary) as well as any other relevant information it has collected under the Directive (Article 10). As described in Section 2.2, the confidentiality provisions of the Directive require that material which qualifies as confidential be kept secret by the Commission and the other Mémber states (Article 11.4). Furthermore, only those authorities responsible for receiving notifications may examine the dossiers transmitted from other countries. In order to ensure that a Member country need not compromise the confidentiality of its industries' trade secrets, Article 11(4) further provides that a Member state which has stricter protection policies for industrial or commercial secrecy than others is not obligated to transfer information to those states which would not enforce the stricter provisions. Article 12 of the Directive, however, states that the competent authority of Member states and the Commission shall have access to the notification dossier and the additional information at all times.

In the United States, EPA may not disclose any information (confidential or not) submitted by a pesticide applicant or registrant under the Federal Insecticide Fungicide and Rodenticide Act (FIFRA), to a person or an agent of a person "engaged in the production, sale or distribution of pesticides in

countries other than the United States or in addition
to the U.S., or to any other person who intends to
deliver such data to such foreign or multinational
business or entity" [Section 10(g), FIFRA as
amended]. EPA is also required to maintain a list of
all persons to whom such data are disclosed and to ap-
prise the submitter of their names and affiliations.
This provision was introduced because the Congress was
concerned about the impact on international competition of using data overseas where compensation
provisions would not apply.

In France Law No. 80-538 of 16th July, 1980
forbids the communication abroad of certain commercial
information. French nationals or residents, or
employees or agents of companies in France, may not
provide foreign authorities with documents or information on economy, trade industry, finance or
technical matters that would "harm the sovereignty,
the security, the essential economic interests of
France or its law and order". These national interests may be specified by the administrative
authorities as needed.

At present, even some exchange of
non-confidential information may pose some difficulties. For the time being no authority exists for
this exchange, and in some countries the disclosure
could be precluded by general principles governing
administrative secrecy (e.g. in Japan, Belgium and
Italy). However, in Canada, Finland and the United
States for example, no problems are anticipated for
the international exchange of non-confidential
information.

A few countries, such as France, (Law
No. 80-538 of 16th July, 1980), have explicitly
reserved an exception to restrictions on information
being transmitted abroad where the country is party to
an international agreement.

PART II


CONFIDENTIALITY OF DATA AND PROTECTION OF
PROPRIETARY RIGHTS IN RELATION TO THE DISCLOSURE
AND THE EXCHANGE OF INFORMATION OF CHEMICALS


Final Report of the Group of Experts
on Confidentiality of Data

## ORIGIN AND PURPOSE OF THE WORK

Satisfactory control of chemicals for the protection of man and the environment necessitates the
transmission of information. This necessity arises
from the aim of achieving a harmonized approach to
hazard assessment of chemicals within the OECD Member
Countries and also an international sharing of efforts. Confidentiality of data can obviously be an
obstacle to the transmission of information. The obstacle can arise at the national level and, even more
obviously, at the international level. Another consideration is, that if data are used differently by
different countries, confidentiality could lead to
restriction or even actual barriers to trade.
Understandably, therefore, the question of confidentiality was one of the important issues addressed
in a special programme proposed by the 16th Meeting of
the Chemicals Group of the Environment Committee
(18th-20th April, 1978) in response to the conclusions of the International Meeting in Stockholm
(11-13th April, 1978). The proposal was welcomed by
the Environment Committee at its 22nd Meeting
(24th-26th April, 1978) and led to a Council Decision
[C(78)127(Final)] establishing a Special Programme
covering the four priority areas defined at Stockholm,
including confidentiality of data. In Article 1 of
Part II of its Decision, the Council stated that the
purposes were as follows:

- to contribute to the protection of man and his
  environment from chemicals' hazards;

- to prevent the creation of non-tariff barriers to trade (arising from discriminatory or
  cumbersome administrative treatment, if only
  through proliferation of formalities at
  international level).

The Decision provided that the Management
Committee responsible for implementing the programme
could propose changes to it. Provision was also made
for high level meetings of the Chemicals Group to give
general orientation to the work as the need arose. At
its meeting, the Management Committee accepted an
offer by France to act as Lead Country for the Group
of Experts on the Confidentiality of Data. An annex
to the Decision laid down the broad directions and
aims of the activity (Annex I).

DEVELOPMENT OF THE PROGRAMME
_____

        A  work  programme  adopted  by  the  Group  of
Experts  at  its  first  meeting  was  approved  by  the
Management Committee.  The plan was to start by circulating a questionnaire, to collect information about
current practice in Member countries on confidential
data  protection,  particularly  in  the  context  of
chemical control regulations.

        The questionnaire was compiled at the Group's
first meeting, and was quickly circulated.  The replies were analysed in two reports.  These highlighted
the  sharp  contrast  between  countries  in  which  administrative secrecy is an absolute rule and those in
which  public  access  to  administrative  documents  is
guaranteed by law.  In the former, information submitted to the government is covered by administrative
secrecy and cannot be disclosed.  In the latter, information  submitted  to  the  government  may  be  disclosed, unless it fulfils certain conditions making it
possible to prohibit public access.  The replies have
also shown that in most countries national legislation
does not deal specifically with the issue of the international transmission of data.  In many countries,
however, general provisions on confidentiality and administrative secrecy would preclude the transmission
of confidential data on chemicals to other countries
in  the  absence  of  overriding  international
agreements.  Generally speaking, the questionnaires
had been completed very fully indeed, clearly bringing
out  the  significant  differences  from  country  to
country, from whether or not they had chemical control
legislation to different definitions of industrial and
commercial  secrecy  and  different  ways  of  assessing
confidentiality, especially for health and safety data
(and as mentioned above, differences in public acces
to  information).  The  differences  strikingly  showed
what  difficulties  might  arise  in  exchanging  confidential data on a large scale, and how useful it was
to open up some discussion.

        At  all  events,  the  replies  to  the  ques-
tionnaire, and the unique combination of information
elicited by the Group's work seemed of such interest
as to warrant publication in Part I of this book.

        Aware of the difficulties of exchanging con-
fidential data, the Group undertook at its second
meeting, as part of its original plan of work, to

identify data which could be circulated without rais-
ing any confidentiality problem, and which countries
ought in principle to be able to exchange. The MPD
set, which lists the elements necessary for a first
hazard assessment of a chemical, some of whose com-
ponent data appeared not to be confidential, served as
a guide for work.

At the same time, the Group undertook to study
the problem set out in points 5 and 6 of the Mandate,
relating to the consequences of data exchange for confidentiality, and the scope of such exchange. At its
third meeting, the Group set out to elaborate a series
of principles to govern data exchange to provide a
basis for an international agreement on data exchange,
of which the details remained to be defined.

The Group also considered point 8 of its
Mandate, which raised the problem of protecting proprietary rights. Most OECD Member countries have
enacted or are contemplating chemical control
regulations obliging manufacturers and importers to
provide the national authorities with various kinds of
data, including some based on laboratory tests, with
which to assess the effects of a chemical on man and
the environment. Much of this information may be of
commercial value and its disclosure to the public and
competitors might harm the competitive position of the
company reporting it to a national authority. The
question is of particular importance when an international exchange of the information is to be
considered.

An interim report from the Group's Chairman to
the High Level Meeting of the Chemicals Group on 19th
to 21st May, 1980, showed that different national approaches to the problems of confidentiality, disclosure, and the safeguarding of proprietary rights
would make the essential task of harmonizing
regulations on the control of chemicals more difficult. It also showed how differences in attitudes
and methods, deeply rooted in the thinking and legal
systems of the different countries, were just one as-
pect of something far broader than chemicals control.
It mentioned three areas in which progress by the
Group could enlarge, in the short or long term, the
possibilities for exchanging information:

-            definition of the areas of data which
             are not confidential. The transmission
             of these data would not present great
             difficulty and few preconditions would
             be necessary even though some obstacles
             might still exist in certain countries;

-       development of mechanisms for relatively
        straightforward cases whereby
        information might be exchanged between
        governments while the rights of
        companies were protected and the public
        informed;

-       definition of principles concerning the
        exchange of confidential information,
        for possible future action by the OECD
        in a form yet to be defined.

This direction of the work was well received by
the High Level Meeting and reinforced by its conclusions as follows:

"The High Level Meeting:

(i)     affirmed that the exchange of in-
        formation and health and safety data
        between Member countries is necessary
        for purposes of assessment and other
        uses relating to the protection of man
        and the environment;

(ii)    requested that the Expert Group on the
        Confidentiality of Data elaborate those
        principles which would be followed in
        the exchange between governments of con-
        fidential data;

(iii)   requested that this Expert Group develop
        proposals for an international ar-
        rangement or instrument which would
        allow for the transmission of data and
        information between governments, while
        providing appropriate safeguards for
        commercially sensitive data and pro-
        tecting their value".

The Group subsequently concentrated on the aims
assigned by the HLM guidelines, studying various aspects of the problem further.

The Group's discussions on data not needing
protection resulted in a list and recommendations [see
Conclusions, (ii)].

In order to permit exchange of confidential
data in spite of the difficulties arising from
differences in the way governments assess the

confidentiality of data submitted to them under
national regulations, and differences in the way some
of them are disclosed, the Group undertook to define
principles which, if respected by Member countries,
might in present circumstances permit such exchange
between national authorities. In defining the principles, considerable account was taken of the need to
protect the proprietary rights of the data exchanged
[see Conclusions, (iii)].

The safeguarding of proprietary rights was also
the subject of a proposal which, though limited, would
be capable of further development [see Conclusions,
(iv)].

The Group tackled a number of problems associated with the exchange of confidential data,
especially economic problems, the duration of confidentiality, practical arrangements for exchanging
information, and what international instrument
(mechanism) it might be able to propose for international exchange.

The Group then decided to draw up a com-
prehensive report on its work, mentioning not just its
conclusions but also the difficulties, disagreements,
negative conclusions and unsolved problems, so as to
provide a maximum of inputs for further progress. The
present document is designed to provide such a full
report.

## FIELD OF APPLICATION OF THE GROUP'S WORK - SOME BASIC PROBLEMS ASSOCIATED WITH THE EXCHANGE OF DATA

The Group considered that it should be concerned exclusively with data for chemicals in their
relationship with the protection of man and the environment, meaning all such data brought to the knowledge of national authorities, for existing and new
chemicals, and especially data collected:

- under notification procedures of new
  products, and procedures for testing or
  notifying hazards of existing chemicals;

- in studies carried out directly under the
  sponsorship of government agencies.

The Group, while addressing the issue of
chemicals in general, was aware that certain
categories of chemicals, like pesticides,

pharmaceuticals, food additives, etc., may be dealt with in a manner more specifically related to the category.

The control of chemicals involves transmitting data. Data may be exchanged or transmitted among several partners: from companies to the responsible authority of a country's administration, from the administration to the public, between different authorities in the administration of any one country, and lastly, between the responsible authorities in different countries.

The Group did not feel competent to consider the transmissions, inside a country, of data received under national regulations. That was a matter for legislators in the individual countries. A chemicals regulating authority exists only because of the need to protect citizens and their environment against the harmful effects of certain products. It is up to the authority to strike a fair balance between the need to inform the public and the need to protect, in accordance with national legislation, the confidential character of data collected domestically.

The Group discussed whether a time limit should apply to the confidential treatment of data. The Group recognized that every item of information loses its confidential character after some length of time, but agreed that the latter can only be assessed individually for each case. Hence it concluded that no general rule can be laid down and that the responsible authority of a Member country has to decide on the issue within the framework of its legislation, if any.

The Group did try to evaluate the possibility of exchanging information among different countries. Exchange of confidential information between countries is not yet commonplace, and the issue is generally not addressed in national legislation. General provisions on confidentiality and administrative secrecy applying in many Member countries, however, presently preclude the transmission of confidential data on chemicals to other countries. When such countries wish to exchange secret or confidential data, they sign conventions empowering them to do so.

It has been emphasized, that the exchange of data was not intended to require the development of new data; it was available data that would be exchanged. One of the essential points was to make best use of the available data and avoid duplication of testing, in view of the world scarcity of facilities and intellectual resources for conducting tests. The use of such data does however raise economic and financial problems, which will be mentioned in the continuation of the report.

The Group has recognised that this exchange of data between national authorities should complement the information received through the normal flow of information between companies and governments. The exchange between countries should not distort competition and, for example, should not exempt companies from normal testing and reporting requirements for the marketing of a new chemical. Neither should it preclude those companies who so wish from dealing directly with foreign authorities.

At present, countries differ in the extent of legislation on the control of chemicals and consequently on the ability to generate and collect data and protect them with a proper degree of confidentiality. There may therefore be an imbalance in the advantage different countries gain from the exchange of information: countries with well developed legislation might be giving information to countries with little to offer in return. This imbalance will become less pronounced if legislation becomes more fully harmonized. Some members of the Group believe that, in the meantime, a country should have appropriate legislation, or, at least, a firm commitment to such legislation before requesting information from other countries and that such a condition, by redressing the imbalance and, at the same time, offering greater assurance of confidentiality, would facilitate rather than impede the exchange of information.

The Group's discussions showed a broad consensus on the non-confidential character of certain data and on the absolutely confidential nature of other data, such as production data. As to another category, health and safety data, of particular relevance to the protection of man and the environment, countries differ in the status they attribute, some having legislation of public access to such information and organising disclosure, sometimes not without certain precautions, whereas others regard them as absolutely confidential. The Group recognised that no common position on this category could be reached since differing disclosure policies were deeply rooted in national political systems which it would be unrealistic to expect to change rapidly. Therefore, such data can probably be exchanged, in present circumstances, only if countries agree to adhere to principles which would in fact guarantee that the confidential character of data would be protected whenever the country providing them so required.

Because of the lack of harmonization in chemicals control, and in assessing the confidentiality of certain kinds of data, broad exchange of confidential data must wait upon greater harmonization of chemical control procedures and of data disclosure practices. Exchanging confidential data, which the Group seeks to encourage, could only facilitate harmonization and would not bring it about; some members of the Group have urged that its proposals for facilitating the exchange of confidential data should be part of a broader effort to harmonize chemical controls procedures, and not brought in as a substitute for harmonization. In their view, the proposals should help to encourage Member countries who have not yet enacted legislation on the question to do so during the forthcoming years.

## NON-CONFIDENTIAL DATA

The Group felt that developing a consensus on a list of non-confidential data was in fact the initial step in addressing the High Level Meeting's first recommendation relating to exchange of information for purposes of hazard assessment of chemicals and other uses relating to the protection of man and the environment. Such a list could be widely circulated with no adverse effect on the companies which generated the data.

Some of the data required for assessment are made available by companies to their customers for the protection of workers handling the chemicals. Others are circulated for environmental protection purposes. Since this information receives a certain degree of publicity, there is no reason why it cannot be freely exchanged.

Meanwhile, another OECD Group of Experts established a list of properties and data the knowledge of which they consider essential for a preliminary assessment of the potential effects of a new chemical on man and the environment (MPD). The Group thought it sensible to use this as a starting point in compiling a list of data which could be circulated and exchanged without restriction.

In the list of non-confidential data, the Group has not included the chemical identity of a substance or the composition of a product, for although that

information is generally considered essential for assessing a chemical, it is often claimed confidential
by companies marketing new chemicals. Whereas certain
existing procedures for new chemicals, e.g. TSCA
Section 5, may recognize such a claim, other procedures contain provisions for mandatory disclosure of
chemical identity, e.g. Council Directive 79/831/EEC
which states that the identity of a new substance
classified dangerous within the meaning of the
Directive must be disclosed. The Group, therefore,
thought that this issue could be addressed in a more
appropriate way in the context of harmonisation of
notification procedures.

There should probably be no difficulty about
circulating the kind of data mentioned in the list
among the various government departments or elsewhere
inside a country. It is expected also that the international transmission of such data would not be op-
posed by existing national legislation which is
generally silent on the issue. If an obstacle of this
nature would arise it could certainly be removed by an
agreement between countries.

The list of non-confidential data, together
with a preamble briefly summarizing the ideas
mentioned above and shedding some light on the scope
which the Group feels able to give it can be found
under Conclusions, (ii).


EXCHANGE OF CONFIDENTIAL DATA

Having produced a consensus on a general rule
of non-confidentiality for certain data, the Group
then concentrated on the transmission of confidential
data between countries. An item of information may be
confidential for any of several reasons: it may
directly reveal facts whose secrecy is vital for the
company, e.g. manufacturing processes, undisclosed
know-how, certain marketing aspects such as a list of
customers, the product's profit margin, etc. Such
data are not usually disclosed outside the firm, and
are deemed highly confidential when made available to
anyone outside the firm, like government officials.
All countries have laws or regulations defining the
concept of "business secrecy", "industrial and commercial secrecy", "manufacturing secrecy" etc.
Furthermore, data transmitted to the authorities for
statistical purposes are more specifically covered by
statistical secrecy. Therefore, the various kinds of
secret cannot be transmitted from one state to another
without special protection.

Data submitted by a company to a national control authority will include some which, though not really business secrets, the company would prefer not to have divulged. These are health and safety data to which the company will have had to devote considerable amounts of time and resources. Companies, when obliged to submit this kind of data in view of entering the market, will tend to claim confidentiality to protect their proprietary rights and competitive position. Disclosure of the data to competitors could result in their entering markets without the same expenditure of time and resources. The government should allow for this in balancing the need to protect property rights against the need to inform its citizens. A government receiving this kind of information usually recognises it as being confidential to some degree. However, some countries have regulations making exceptions to this rule. One instance among others is the United States, where health and safety data can be disclosed under Section 14 of the Toxic Substances Control Act and Section 10 of the Federal Insecticide, Fungicide, and Rodenticide Act.

The Group believes that exchange of confidential data on chemicals requires multilateral or bilateral agreements. Such agreements should contain a number of conditions or principles, and the Group has sought to identify these. While it has not been unanimous in defining them, the Group did agree on the following essential basis:

- respect for the sovereignty of the country transmitting information in its decision on the confidential nature of information

- respect for proprietary rights in data

- respect for the free play of competition

The principles are mentioned in Conclusions, (iii). They are accompanied by comments reflecting the various opinions expressed in the discussion.

Once such principles have been accepted, the Group considers that OECD should encourage Member countries to agree by appropriate means to supply such information, especially health and safety data, as may be requested by other countries.

## FINANCIAL CONSEQUENCES TO INDUSTRY OF DISCLOSURE OF DATA

The Group several times approached the problem of evaluating the financial losses to industry of the lawful or accidental disclosure of certain data. Members of the Group were several times invited to contribute material towards an answer to the question. Two Delegations did so, and a bibliograhical search yielded two studies carried out in the United States. But the material the Group collected was very general and qualitative. Specific information and figures are lacking, for two reasons: financial losses can only be evaluated case by case, in terms of the advancement and cost of the research already undertaken and the potential commercial value of the product for which information had been disclosed; and, in the event of disclosure, or illegal use, of data by a company, the amount of loss to the company owning the data is usually itself strictly confidential.

## THE PROTECTION OF PROPRIETARY RIGHTS AND PROBLEMS OF COMPENSATION

The Group received two requests to consider protection of proprietary rights and compensation problems. The Group's mandate requested it to be aware of the implications of disclosure of data for testing costs and their reimbursement (Annex I). The High Level Meeting more specifically requested the Group to put forward proposals for an international arrangement or instrument under which data could be exchanged, including measures to protect the value of commercially sensitive data.

Some countries are under a legal obligation to disclose health and safety data, and countries may have a need to disclose such data in order to provide adequate information to the public. Disclosure of the data may however affect the competitive position of the company which submitted them.

The dilemma could be considerably simplified if the emphasis were put on the use of data rather than on their disclosure. A company submitting health and safety data very often insists that their confidential nature be respected, not because of their value as business secrets, but because the company wishes to prevent its competitors from appropriating them and using them for purposes such as approval, registration or notification. The confidentiality clause would be less necessary if exclusive use of the data submitted were the general rule.

It appears difficult to apply the exclusive use principle internationally in the present state of harmonization of control provisions and requirements for data submission imposed on companies. One of the obstacles is the disparity between notification systems already in place, some requiring only one notification of a new product, made by the first producer, others requiring notification by all succesive producers of the same new product. Another obstacle is that some governments use all available data for assessment without distinction as to owner, although some impose a compensation system.

With respect to notification of new chemicals a particular problem arises when health and safety data submitted by a notifier are disclosed to the public. Subsequent notifiers in the same country or notifiers in a different country may be able to use the data for their notifications and enter a market without the same expenditure of time and resources, thus gaining an unfair advantage over the first notifier.

The Group formulated the proposal set out in Annex IV. By requiring a notifier to show that the data were actually his property, or that he had obtained the agreement of the owner to submit them, original data would be protected from imitation and copying. This proposal would solve only a part of the problem of proprietary rights. It could nevertheless constitute a first stage and, if it proved satisfactory, lead on to broader protection systems. The proposal may also facilitate compensation between firms, which in the opinion of the Group should be the subject of private agreements with the minimum of government involvement.


MECHANISMS FOR EXCHANGE

The Group, reflecting on the mechanisms for exchange, did not propose routine exchanges, and in this made no distinction between confidential and non-confidential data. Respect for confidentiality would be most effectively guaranteed if exchanges were by request, between the responsible authorities in the requesting and supplying countries. For the same reason, the Group did not like the idea of a central clearinghouse with a data bank.

The Group recognised that direct communication between the responsible authorities would not work when a government did not know to which country it should apply for information. The Group is divided on whether it would be appropriate to have a central point not, this time, relying on a data bank but operating simply as an office to channel requests. To be effectual, such an office would nevertheless have to possess a register of the data available in the different countries, which would in turn have to compile lists and keep them up-to-date.

## THE INTERNATIONAL INSTRUMENT

The Group was invited by the High Level Meeting of the Chemicals Group to formulate proposals for an international instrument for exchange of data among Member countries. The experts had different views as to how this request should be met, some considering that the Group lacked the necessary legal and political competence. The Group then agreed that it rather would highlight in annexes to the report, the outcome of its discussions on three important issues: non-confidential data, exchange of confidential data and protection of proprietary rights.

The Group recommends that the Management Committee carefully consider Conclusions (ii), (iii) and (iv) and decide on appropriate actions to be taken within the OECD. While the Group is not prepared to recommend the specific type of OECD action, such as a Council Decision or Recommendation, it suggests a differentiated approach in relation to each issue.

With respect to Conclusions (ii), Non-Confidential Data, the Group in its majority recommends some type of multilateral agreement by which OECD Member countries would agree to the principle of full disclosure and exchange of this type of data. Because this proposal on non-confidential data is not controversial a multilateral agreement would have inherent advantages for achieving a full exchange within OECD.

With respect to Conclusions (iii), Principles for Exchange of Confidential Informaton, the Group recommends that exchanges be accomplished first through bilateral agreements between Member countries. Because the principles proposed by the Group are general, it would be best to try such exchanges bilaterally and use the bilateral exchanges as a means of testing and refining the principles.

The Group further recommends that, when the principles have been tested and refined, multilateral agreements be entered.

With respect to Conclusions (iv), Protection of Proprietary Rights of Data, the Group recommends some type of multilateral agreement by which all Member countries would provide this protection of proprietary rights. Only a multilateral agreement would retain the validity of the concept and reduce possible adverse commercial effects due to the disclosure of complete health and safety data.

## CONCLUSION

### (i)    General

Pursuing the twofold aim of improving the protection of man and the environment, and reducing barriers to international trade, the Expert Group on the Confidentiality of Data feels that it has gone as far as is now possible in a very delicate area, impinging as it sometimes does on deeply-rooted national traditions. The Group has been anxious not to go outside its competence, and leaves it to higher authorities, assisted by lawyers, to determine what actions can be taken at the international level. The Group has however indicated its preferences with regard to the appropriate approaches of three key issues: non-confidential data, exchange of confidential data, and protection of proprietary rights. The Group, extensively discussing in the body of the report the difficulties it encountered, does not intend to slow down or discourage possible action. The Group rather does want to show clearly how much of the road remains ahead and where work is needed.

The Group is convinced that momentum will be generated by the first exchanges, for which it has sought to lay the foundations, and that, with experience, this will help gradually to eliminate the remaining obstacles to their large-scale development. Thus, while it has not yet been possible to envisage an open system for exchanging health and safety data, together with a sufficiently reliable international mechanism for the protection of proprietary rights, a limited mechanism, applicable in notifications of new chemicals, has been proposed. If this mechanism proves satisfactory, in application, the way will be open for future extensions to it.

The Group is also persuaded that a balanced and equitable distribution will come about among the various OECD Member countries as regards studies of the effects of chemicals on man and the environment, without the need for complicated financial compensation systems. These, the Group considered, would present practical difficulties at international level. The Group thereby did not rule out inter-company reimbursements to obviate, in appropriate cases, duplication of testing. The States might even facilitate those without intervening in any official, constraining way.

The Group is convinced that its proposals, far from eliminating the need for the current OECD efforts to harmonize chemical control regulations, will on the contrary facilitate those efforts, harmonization being a prerequisite, in the medium term, for the development of the exchange of confidential data and the protection of proprietary rights.

The Group considers that the effort to harmonize regulations should also be pursued in the area in which it has worked for the past three years, particularly as regards assessment of the confidentiality of health and safety data and their disclosure to the public. The Group is aware that this will be a long-term task, having found that it could not make much progress along these lines during its short period of activity. It hopes however to have clearly drawn attention to the importance of this issue, since harmonization here would greatly facilitate the exchange of health and safety data and hence, harmonization in chemical product regulation at international level.

The Group recognises that it has given no precise answer to two types of question on which short-term studies should be pursued elsewhere:

- economic problems (in the broad sense, not only those with a direct financial impact) arising out of accidental or controlled disclosure of confidential data. These issues are connected with those currently being studied under the economic programme of the Chemicals Group (Programme on the Economic and Trade Effects of Chemicals Control);

- practical ways available to countries to gain
knowledge of the existence of data which can
be exchanged under present circumstances,
given that members of the Group did not reach
consensus on the desirability of a central
clearinghouse. Some experts proposed the
keeping of a central index listing chemicals,
by identity or trade name, together with the
names of the countries possessing data to
guide countries in their efforts to obtain
data. The need to address this question is
more urgent with respect to exchange of non-
confidential data, a possible short term
objective, than with respect to exchange of
confidential data.

Despite these gaps, the Group feels that it has
made an appreciable contribution to the development of
the exchange of data, including confidential data,
among Member countries. It hopes that measures will
be taken as soon as possible at the appropriate level,
to encourage Member countries to develop the exchange
of non-confidential data, to develop that of confidential data by applying principles permitting this,
and, although this would be a more limited action, to
apply the measures recommended for recognising the
right of notifiers to use data.

(ii)   Non-Confidential Data

Preamble

The Group believes that certain data, of value
for hazard assessment of chemicals and other purposes
connected with the protection of man and the environment, may be termed non-confidential, and it pro-
poses the following list of such data.

In this context, "non-confidential" means that
no restrictions should be put on the exchange of the
data between governments nor on the free access of the
public to such data.

The Group recommends that Member countries take
the necessary steps to enable governments to exchange
and disclose the data when available. The Group is of
the opinion that the data should be exchanged between
governments on request and not as a matter of routine.

The Group does not mean the list to be restrictive. The Group recognises, on the contrary,
that in some circumstances there may be other data

which are considered non-confidential both by the government and the submitter and believes that if these are useful for hazard asessment of chemicals, they should also be exchanged. The list below is in-spired by the OECD Minimum Pre-marketing Set of Data, but is not meant to be restricted to information on new chemicals. The quality of non-confidentiality, as defined above, of the data listed also applies to existing chemicals, and therefore to all chemicals.

The List of Non-Confidential Data

- trade name(s) or name(s) commonly used

- general data on uses (the uses need to be described only broadly, like: closed or open system, agriculture, domestic use, etc.)

- safe handling precautions to be observed in the manufacture, storage, transport and use of the chemical

- recommended methods for disposal and elimination

- safety measures in case of an accident

- physical and chemical data with the exception of data revealing the chemical identity (e.g. spectra). If the physical and chemical data make it possible to deduce therefrom the chemical identity only ranges of values need be given

- summaries of health and safety data including precise figures and interpretations. (The submitter of the health and safety data should participate in the preparation of the summaries).

(iii)   Suggested Principles to govern the Exchange of Confidential Information between Member Countries

Preamble

1.      The Chemicals Group at its High Level Meeting in May 1980 stated that the exchange of health and safety data between Member countries was necessary for the purpose of assessing chemicals with the object of protecting man and the environment. It instructed the

Group of Experts on the Confidentiality of Data to work out the principles applicable to the exchange of confidential data.

2.     The Group of Experts, in its discussions, defined the scope of such exchange;   it was agreed that the exchange of information between the authorities in Member countries responsible for the control of chemicals should complement the company submissions to these authorities and secondly should allow for exchange on request when companies are not involved. Given the worldwide scarcity of material and intellectual resources for conducting tests, exchange, avoiding duplication of tests so far as possible, should enable better use to be made of existing data.     Exchangeable data should be both for new chemicals and for existing chemicals.

3.     Member countries differ very widely in their assessment of the confidentiality of data submitted in response to regulations or administrative practice for chemicals control.  While it is generally recognised that the notifier is entitled to claim confidentiality for some of the data he makes available to a competent authority,   the   final   decision   lies   with   the authority. As a result certain data, which cannot be disclosed in some countries, may be disclosable in others.  The extent to which confidential data are circulated within government departments may also vary from one country to another.  The exchange of data between countries therefore raises a problem of widening the access to confidential data.

       Confidentiality of data is certainly the factor most often limiting exchange of information on chemicals between countries.  The Group therefore considered it opportune to recommend that certain types of data should not be designated as confidential [see (ii) above] and that their exchange should not be limited by principles.

4.     It would seem to be premature, at the present stage, to try to solve these problems by proposals aiming at international harmonization of the relevant laws. Even if greater harmonization of chemicals control regulations can be achieved, the purpose underlying the work of the OECD chemicals programme, the fact remains that concepts of administrative secrecy, and of industrial and commercial secrecy, in different countries derive from fundamental principles associated with national law, which must act as a curb on harmonization.  The Group has pointed out that the OECD work towards harmonization should, in particular,

encourage those Member countries which have not yet adopted legislation on chemicals control to do so in the years ahead.

5.      The exchange of confidential data between countries should be governed by principles taking account of the differences between legislation and administrative practice in different countries, and enable countries to participate in such exchange without infringing in the law or practice prevailing in their own territories.  Clearly, a list of principles extensively respecting the traditions of countries strictly applying the rule of administrative secrecy to any information imparted to the government imposes restrictions on the possibilities for exchange.  A competent authority will only transmit confidential information if it can be certain that the requesting authority will treat it at least with the same degree of confidentiality as is practised in the transmitting country.  Countries whose laws or administrative practices favour disclosure could agree to follow less restrictive principles in a mutual exchange of data.

6.      The principles were defined by the Group on the following basis:

- the exchange system must respect the sovereignty of the country transmitting information in its decision on the confidential nature of the information;

- a competent authority must make every reasonable effort to obtain the information available in its country before requesting confidential information from the competent authority in another country;

- exchanges of confidential data between competent authorities in different countries should not distort competition and in particular, should not have the effect:

- of subjecting nationals in the solicited country to a more severe testing or reporting requirement than would apply to a national of the soliciting country in the same situation;

- or exempting nationals of the soliciting country from conforming to the notification procedures prevailing in their country;

- all data made available to a competent
authority must remain the property of the
submitter, even after exchange with competent
authorities elsewhere, to the extent recognised in the original country.


The text of the principles drawn up by the
Group is presented below, accompanied by explanatory
comments reflecting the various opinions expresed
during the work of the Group.


## PRINCIPLE NO. 1

THE EXCHANGE OF CONFIDENTIAL INFORMATION ON CHEMICALS
BETWEEN THE COMPETENT AUTHORITIES OF COUNTRIES IS
INTENDED SOLELY TO FACILITATE THE HAZARD
ASSESSMENT OF CHEMICALS AND THE
PROTECTION OF MAN AND THE
ENVIRONMENT


## Comments

7.      The Group distinguished three categories of
confidential information that might be available to a
competent authority and might be exchangeable between
Member countries:    data reported under chemical
control legislation or regulation, or in the normal
course of chemical control administration, data
supplied by companies voluntarily or upon request, and
data produced under the sponsorship of government
departments and other public services.  The Group was
mainly interested in the exchange of data in the first
category, pointing out that such exchange should not
be an alternative to ordinary submissions by companies
to competent authorities.

8.      It seemed difficult if not impossible to
establish principles which could govern the exchange
in the two other categories.  The discretionary power
exercised by the competent authority in deciding or
declining to transmit its own data, or data provided
voluntarily by companies, lends itself to no general
rule and will be different from case to case.
However, there should be nothing to prevent such data
from being exchanged when appropriate.

9.      From the standpoint of protecting man and the
environment, the Group considered that it should not
define the term "chemicals".  The Group also made no
distinction between existing chemicals and new
chemicals.  This distinction becomes very difficult in
an instance where data are exchanged between countries

whose systems of notification of new chemicals are different in scope. For similar reasons, it did not appear desirable to distinguish chemicals in terms of the particular use made of them, and to exclude some categories from eligibility for exchange.

10.    Exchange is intended to transmit data already available to the competent authority, and not to have the transmitting authority gather and develop new data for this purpose.

## PRINCIPLE NO. 2

A COUNTRY HAVING RECEIVED INFORMATION IN RESPONSE
TO A REQUEST MUST IN NO CIRCUMSTANCES USE SUCH
INFORMATION FOR ANY PURPOSE OTHER THAN THE
ASSESSMENT OF HAZARDS OF CHEMICALS AND
THE PROTECTION OF MAN AND THE
ENVIRONMENT

Comments

11.    This limitation of the uses that can be made of information transmitted accurately reflects the need recognised by the Chemicals Group at its May 1980 High Level Meeting.    Any extension of the use of information received would prejudice the smooth running of the exchange and the maintenance of the commitment entered into by the countries participating in it.

## PRINCIPLE NO. 3

A COUNTRY, WHENEVER REQUESTING INFORMATION ABOUT
A CHEMICAL, MUST SUBSTANTIATE THE NEED FOR
THE INFORMATION, ON THE GROUNDS THAT:

A)      THE CHEMICAL IS PRESENT OR IS SHORTLY TO
BE MARKETED IN ITS TERRITORY;   AND

B)      THE INFORMATION IS NECESSARY FOR THE
ASSESSMENT OF ITS HAZARDS AND THE
PROTECTION OF MAN AND THE ENVIRONMENT

Comments

12.    Automatic exchange of the available data among all Member countries would be an administrative burden and is not considered worthwhile.    Such an exchange

would also increase the risk of disclosure of confidential data. Therefore, data would be exchanged only in response to a substantiated request.

13.    Linking the acceptability of a request to the information needs as defined in the principle helps to avoid excessively frequent requests, making exchange impractical, and to avoid undue latitude in the reasons a country can give for declining it.

14.    The expression "present ... in its territory" has been chosen to include not only the presence of a chemical on a country's market but also its presence in the country's territory due to transfrontier pollution. The expression "shortly to be marketed" was chosen to include chemicals for which the marketing process has been launched even though the chemical is not yet physically present in the territory.

15.    Several experts considered that the principle above would be too restrictive and reduce the value of the exchange of information in respect to hazard assessment.    They    suggested    supplementing    the principle by:

> "OR    DEMONSTRATE    THE    USEFULNESS    OF    THE INFORMATION    BECAUSE    OF    A    SIMILARITY    IN STRUCTURE TO A CHEMICAL PRESENT OR SHORTLY TO BE MARKETED IN ITS TERRITORY"

However, other experts were of the opinion that the present state of scientific knowledge does not allow the establishment of a direct relationship between chemical structure and effect upon man and the environment which can be generally applied.    Those experts also thought that the concept could harm the proprietary rights of a manufacturer of a chemical showing "similarity in structure" without its chemical being directly concerned or relevant to the case under consideration.

The Group agreed that Member countries could include a provision on structural similarity in bilateral exchange agreements.

## PRINCIPLE NO. 4

### A COUNTRY REQUESTING INFORMATION

A)    MUST ABIDE BY THE DECISION MADE BY THE TRANSMITTING COUNTRY IN RESPECT TO THE CONFIDENTIAL NATURE OF THE INFORMATION;

B) MUST TREAT THE TRANSMITTED INFORMATION WITH AT LEAST THE SAME DEGREE OF CONFIDENTIALITY AS IS PRACTISED IN THE COUNTRY FROM WHICH THE INFORMATION HAS BEEN REQUESTED

C) MAY MAKE THE INFORMATION AVAILABLE TO NATIONAL, REGIONAL OR LOCAL AUTHORITIES ONLY WHEN NECESSARY FOR PURPOSES OF HAZARD ASSESSMENT OF CHEMICALS OR PROTECTION OF MAN AND THE ENVIRONMENT AND ONLY WHEN SUCH AUTHORITIES ARE ABLE TO GUARANTEE THE SAME LEVEL OF CONFIDENTIAL TREATMENT;

D) SHALL NOT TRANSMIT THE INFORMATION RECEIVED TO ANY OTHER COUNTRY

## Comments

16. The national authority having recognised the confidentiality of information submitted to it has the first responsibility for ensuring that it is effectively safeguarded. The authority can only transmit such information if it is certain that the requesting country will respect the confidentiality of such information.

17. "Treat the transmitted information with at least the same degree of confidentiality as is practised in the country from which the information has been requested" means that the requesting country must treat the information in a manner that is the practical equivalent of the treatment of that information in the originating country. The Group understands that receiving countries will not have legisltion identical to that in originating countries.

18. The Group recognised that different authorities within a country's government may need access to information, and that to make it accessible only to one competent authority would remove much of the value of an exchange of confidential information.

19. Each country should designate an authority to be responsible for transmitting confidential data to another country. The receiving country shall not transmit them elsewhere.

## PRINCIPLE NO. 5

### THE REQUESTING COUNTRY SHALL NOT ASK FOR THE TRANSMISSION OF CONFIDENTIAL INFORMATION WHICH IT DOES NOT HAVE THE AUTHORITY TO COLLECT AND USE UNDER ITS LEGISLATION OR IN THE NORMAL COURSE OF ITS ADMINISTRATION

Comment

20.    Exchangeable information would essentially be limited to data submitted under laws, regulations and practice of control of chemicals. It is therefore necessary to avoid a situation in which countries with stricter notification requirements than others find themselves constantly being asked to provide data.

21.    OECD work under the chemicals programme, and especially work on exchanging confidential data, should be part of a broader effort to harmonize chemicals control procedures, and not be allowed to act as a substitute for harmonization. In particular, it should encourage Member countries which have not yet adopted legislation on the matter to do so over the coming years.

## PRINCIPLE NO. 6

### THE SOLICITED COUNTRY SHOULD CONSULT WITH THE PERSON WHO SUBMITTED THE REQUESTED CONFIDENTIAL DATA BEFORE TRANSMITTING THEM

Comment

22.    Since any exchange involves a further risk of disclosure, whose consequences cannot always be fully assessed by the government, it would seem normal to consult the submitter.

23.    However, it should be clearly understood that this is a consultation and that the final decision must be taken by the government, in accordance with national or international provisions.

(iv)    Protection of Proprietary Rights of Data

Preamble

The Group recognised that data submitted to governments were the property of the person obtaining and submitting them or agreeing that they be submitted by another person.

Governments often confirm the confidential character of some of the information they receive. In thus protecting such information from disclosure, they to some extent protect proprietary rights, thereby safeguarding a competitive advantage for the owner of the information. This is so because competitors cannot obtain such information to use in conducting their own administrative formalities. This will generally apply to any information in the business secret category.

It may happen in some countries that data which do not have a business secret character, but have in- volved their notifier in substantial outlays to obtain them, may be disclosed by a government department. This applies to certain health and safety data.

The Group, without pronouncing on the desirability of disclosing this kind of data, con- siders that they do have a competitive value and that there is a need to avoid allowing their disclosure ad- versely to affect their owner. The Group suggests that data submitted should be accompanied by a dec- laration certifying ownership, along the same lines as in the author's copyright system.

## Proposal

"Each Member country must require each notifier of a new chemical to identify the laboratories which produced each of the health and safety data in the notification. If the laboratories were not owned by or otherwise affiliated with the notifier, the notifier must also be required to provide cer- tification that he had the right to use the data. If the authority receiving the notification finds that the notifier was not entitled to use the data, it must not accept the notification. This process does not change the confidential status of health and safety data".

## Comments

This proposal is limited in scope:

- it only concerns the protection of pro- prietary rights in the context of notifications of. new chemicals, whereas the problem also arises for data collected in respect of existing products;

- at present, not all Member countries yet have systems for compulsory notification of new chemicals;

- the proposal does not rely upon any international level control system and may not
present a sufficient degree of guarantee
against fraud.


The proposal is not new. Directive 79/831/CEE
of the Council of the European Communities already
takes this general approach.

The proposal does not settle problems of how
research costs should be shared among notifiers. It
does however open an avenue for compensation among
companies, whose effect would be to reduce
notification costs and permit a rational utilisation
of existing resources. The position of the Group in
connection with the setting of compensation amounts
has always been in favour of transactions with the
minimum intervention by any supervisory authority, a
position it holds even more strongly in the international context.

## DEFINITION OF THE PROGRAMME

[C(78)127(Final) Appendix]

1.      Firstly, a survey should be compiled of the provisions made in existing (or proposed) national and international schemes for the registration and the regulation of toxic chemicals, or other relevant legislation. This will indicate how national governments and international groupings (e.g. EEC) deal with problems of confidentiality at the present time, or in proposed schemes.

2.      There should be clear identification of the different types of data which are under discussion, e.g. the basic dossier of tests provided by chemical manufacturers, health and safety data, risk assessments, etc.

3.      An attempt should be made to identify the weight which each country assigns to the public interest on the one hand and to commercial secrecy on the other hand, in reaching decisions on confidentiality. Suggestions could be presented on how potential conflicts might be resolved.

4.      Having identified or resolved the differences of opinion regarding the balance between data which should remain secret and that which can be released, those studying the problems could consider how the dissemination of information could be managed.

5.      Within each country there may be a need to pass information on toxic chemicals to other government departments, to regional or local authorities etc. This problem should be considered together with its implications for confidentiality. Is there a danger to confidentiality if more than one national register is maintained?

6.      Consideration should also be given to the type of information to be passed on to other countries. Should this include risk assessments and if so, what effect could this have on commercial interests?

7.　　Although not a first priority, those dealing with the questions of confidentiality of information should be aware of the implications for testing costs and their reimbursement, if data are released.

8.　　Finally, consideration should be given as to how industry could participate in specific decisions on disclosure of information, so that the implications for commercial interests are taken into account.

## EXPERT GROUP

| | |
|---|---|
| AUSTRALIA | J. Bell, I. Carruthers, D. Gascoine H. Jitts |
| BELGIUM | J. Bormans, B. Cordier, G. Eliat |
| CANADA | J. Brydon, R. Demayo* |
| DENMARK | O. Jacobsen, E. Lindegaard, H. Sand |
| FINLAND | A.M. Vahakuopus |
| FRANCE | D. Laheyne*, J.P. Michel, J.P. Parenteau* (Chairman), J. Persoz* |
| GERMANY | B. Broecker, K. Günther, A. Lange, P. Lauffer, H. Lindemann, M. Schroeder |
| GREECE | V. Sotiriadou |
| ITALY | T. Garlanda, M. Grassi, L. Turri* |
| JAPAN | H. Hamanaka, H. Imura, M. Kawasaki, M. Kitano, K. Kobayashi, K. Nishikawa, T. Oshima |
| NETHERLANDS | D. de Bruijn*, L. Rinzema*, W. Steemers |
| NORWAY | A. Foyen, V. Lassen |
| SWEDEN | A. Edling, S. Heurgen, I. Kökeritz, R. Lonngren |
| SWITZERLAND | B. Haldimann, H. Hertig, L. Pioda, P. Tobler |
| UNITED KINGDOM | E. Dodds, R. Grainger*, J. Percy-Davis |
| UNITED STATES | E. Berman, R. Bonczek, E. Cull, B. Grossmann, D. Hoinkes, J. Nelson R. Nicholas, G. Paulson, J. Warren |

| | |
|---|---|
| <u>CEC</u> | G. Del Bino, G. Pechovitch, R. Roy, G. Strongylis |
| <u>WHO</u> | G. Vettorazzi |
| <u>OECD</u> | M. Caron, P. Crawford, H. Gassmann, T. Hiraishi, J. Nichols, R. Ruggeri, H. Van Looy, B. Wagner |

---

\*     Attended all meetings

# OECD SALES AGENTS
# DÉPOSITAIRES DES PUBLICATIONS DE L'OCDE

**ARGENTINA – ARGENTINE**
Carlos Hirsch S.R.L., Florida 165, 4° Piso (Galería Guemes)
1333 BUENOS AIRES, Tel. 33.1787.2391 y 30.7122
**AUSTRALIA – AUSTRALIE**
Australia and New Zealand Book Company Pty, Ltd.,
10 Aquatic Drive, Frenchs Forest, N.S.W. 2086
P.O. Box 459, BROOKVALE, N.S.W. 2100
**AUSTRIA – AUTRICHE**
OECD Publications and Information Center
4 Simrockstrasse 5300 BONN. Tel. (0228) 21.60.45
Local Agent/Agent local :
Gerold and Co., Graben 31, WIEN 1. Tel. 52.22.35
**BELGIUM – BELGIQUE**
LCLS
35, avenue de Stalingrad, 1000 BRUXELLES. Tel. 02.512.89.74
**BRAZIL – BRÉSIL**
Mestre Jou S.A., Rua Guaipa 518,
Caixa Postal 24090, 05089 SAO PAULO 10. Tel. 261.1920
Rua Senador Dantas 19 s/205-6, RIO DE JANEIRO GB.
Tel. 232.07.32
**CANADA**
Renouf Publishing Company Limited,
2182 St. Catherine Street West,
MONTRÉAL, Que. H3H 1M7. Tel. (514)937.3519
OTTAWA, Ont. K1P 5A6, 61 Sparks Street
**DENMARK – DANEMARK**
Munksgaard Export and Subscription Service
35, Nørre Søgade
DK 1370 KØBENHAVN K. Tel. +45.1.12.85.70
**FINLAND – FINLANDE**
Akateeminen Kirjakauppa
Keskuskatu 1, 00100 HELSINKI 10. Tel. 65.11.22
**FRANCE**
Bureau des Publications de l'OCDE,
2 rue André-Pascal, 75775 PARIS CEDEX 16. Tel. (1) 524.81.67
Principal correspondant :
13602 AIX-EN-PROVENCE : Librairie de l'Université.
Tel. 26.18.08
**GERMANY – ALLEMAGNE**
OECD Publications and Information Center
4 Simrockstrasse 5300 BONN Tel. (0228) 21.60.45
**GREECE – GRÈCE**
Librairie Kauffmann, 28 rue du Stade,
ATHÈNES 132. Tel. 322.21.60
**HONG-KONG**
Government Information Services,
Publications/Sales Section, Baskerville House,
2/F., 22 Ice House Street
**ICELAND – ISLANDE**
Snaebjörn Jónsson and Co., h.f.,
Hafnarstraeti 4 and 9, P.O.B. 1131, REYKJAVIK.
Tel. 13133/14281/11936
**INDIA – INDE**
Oxford Book and Stationery Co. :
NEW DELHI-1, Scindia House. Tel. 45896
CALCUTTA 700016, 17 Park Street. Tel. 240832
**INDONESIA – INDONÉSIE**
PDIN-LIPI, P.O. Box 3065/JKT., JAKARTA, Tel. 583467
**IRELAND – IRLANDE**
TDC Publishers – Library Suppliers
12 North Frederick Street, DUBLIN 1 Tel. 744835-749677
**ITALY – ITALIE**
Libreria Commissionaria Sansoni :
Via Lamarmora 45, 50121 FIRENZE. Tel. 579751/584468
Via Bartolini 29, 20155 MILANO. Tel. 365083
Sub-depositari :
Ugo Tassi
Via A. Farnese 28, 00192 ROMA. Tel. 310590
Editrice e Libreria Herder,
Piazza Montecitorio 120, 00186 ROMA. Tel. 6794628
Costantino Ercolano, Via Generale Orsini 46, 80132 NAPOLI. Tel.
405210
Libreria Hoepli, Via Hoepli 5, 20121 MILANO. Tel. 865446
Libreria Scientifica, Dott. Lucio de Biasio "Aeiou"
Via Meravigli 16, 20123 MILANO Tel. 807679
Libreria Zanichelli
Piazza Galvani 1/A, 40124 Bologna Tel. 237389
Libreria Lattes, Via Garibaldi 3, 10122 TORINO. Tel. 519274
La diffusione delle edizioni OCSE è inoltre assicurata dalle migliori
librerie nelle città più importanti.
**JAPAN – JAPON**
OECD Publications and Information Center,
Landic Akasaka Bldg., 2-3-4 Akasaka,
Minato-ku, TOKYO 107 Tel. 586.2016
**KOREA – CORÉE**
Pan Korea Book Corporation,
P.O. Box n° 101 Kwangwhamun, SÉOUL. Tel. 72.7369

**LEBANON – LIBAN**
Documenta Scientifica/Redico,
Edison Building, Bliss Street, P.O. Box 5641, BEIRUT.
Tel. 354429 – 344425
**MALAYSIA – MALAISIE**
and/et SINGAPORE - SINGAPOUR
University of Malaya Co-operative Bookshop Ltd.
P.O. Box 1127, Jalan Pantai Baru
KUALA LUMPUR. Tel. 51425, 54058, 54361
**THE NETHERLANDS – PAYS-BAS**
Staatsuitgeverij
Verzendboekhandel Chr. Plantijnstraat 1
Postbus 20014
2500 EA S-GRAVENHAGE. Tel. nr. 070.789911
Voor bestellingen: Tel. 070.789208
**NEW ZEALAND – NOUVELLE-ZÉLANDE**
Publications Section,
Government Printing Office Bookshops:
AUCKLAND: Retail Bookshop: 25 Rutland Street,
Mail Orders: 85 Beach Road, Private Bag C.P.O.
HAMILTON: Retail Ward Street,
Mail Orders, P.O. Box 857
WELLINGTON: Retail: Mulgrave Street (Head Office),
Cubacade World Trade Centre
Mail Orders: Private Bag
CHRISTCHURCH: Retail: 159 Hereford Street,
Mail Orders: Private Bag
DUNEDIN: Retail: Princes Street
Mail Order: P.O. Box 1104
**NORWAY – NORVÈGE**
J.G. TANUM A/S Karl Johansgate 43
P.O. Box 1177 Sentrum OSLO 1. Tel. (02) 80.12.60
**PAKISTAN**
Mirza Book Agency, 65 Shahrah Quaid-E-Azam, LAHORE 3.
Tel. 66839
**PHILIPPINES**
National Book Store, Inc.
Library Services Division, P.O. Box 1934, MANILA.
Tel. Nos. 49.43.06 to 09, 40.53.45, 49.45.12
**PORTUGAL**
Livraria Portugal, Rua do Carmo 70-74,
1117 LISBOA CODEX. Tel. 360582/3
**SPAIN – ESPAGNE**
Mundi-Prensa Libros, S.A.
Castelló 37, Apartado 1223, MADRID-1. Tel. 275.46.55
Libreria Bosch, Ronda Universidad 11, BARCELONA 7.
Tel. 317.53.08, 317.53.58
**SWEDEN – SUÈDE**
AB CE Fritzes Kungl Hovbokhandel,
Box 16 356, S 103 27 STH, Regeringsgatan 12,
DS STOCKHOLM. Tel. 08/23.89.00
**SWITZERLAND – SUISSE**
OECD Publications and Information Center
4 Simrockstrasse 5300 BONN. Tel. (0228) 21.60.45
Local Agents/Agents locaux
Librairie Payot, 6 rue Grenus, 1211 GENÈVE 11. Tel. 022.31.89.50
**TAIWAN – FORMOSE**
Good Faith Worldwide Int'l Co., Ltd.
9th floor, No. 118, Sec. 2
Chung Hsiao E. Road
TAIPEI. Tel. 391.7396/391.7397
**THAILAND – THAILANDE**
Suksit Siam Co., Ltd., 1715 Rama IV Rd,
Samyan, BANGKOK 5. Tel. 2511630
**TURKEY – TURQUIE**
Kültur Yayinlari Is-Türk Ltd, Sti.
Atatürk Bulvari No : 77/B
KIZILAY/ANKARA. Tel. 17 02 66
Dolmabahce Cad. No : 29
BESIKTAS/ISTANBUL. Tel. 60 71 88
**UNITED KINGDOM – ROYAUME-UNI**
H.M. Stationery Office, P.O.B. 569,
LONDON SE1 9NH. Tel. 01.928.6977, Ext. 410 or
49 High Holborn, LONDON WC1V 6 HB (personal callers)
Branches at: EDINBURGH, BIRMINGHAM, BRISTOL,
MANCHESTER, BELFAST.
**UNITED STATES OF AMERICA – ÉTATS-UNIS**
OECD Publications and Information Center, Suite 1207,
1750 Pennsylvania Ave., N.W. WASHINGTON, D.C.20006 – 4582
Tel. (202) 724.1857
**VENEZUELA**
Libreria del Este, Avda. F. Miranda 52, Edificio Galipan,
CARACAS 106. Tel. 32.23.01/33.26.04/33.24.73
**YUGOSLAVIA – YOUGOSLAVIE**
Jugoslovenska Knjiga, Terazije 27, P.O.B. 36, BEOGRAD.
Tel. 621.992

Les commandes provenant de pays où l'OCDE n'a pas encore désigné de dépositaire peuvent être adressées à :
OCDE, Bureau des Publications, 2, rue André-Pascal, 75775 PARIS CEDEX 16.

Orders and inquiries from countries where sales agents have not yet been appointed may be sent to:
OECD, Publications Office, 2 rue André-Pascal, 75775 PARIS CEDEX 16.

65579-9-1982

OECD PUBLICATIONS, 2, rue André-Pascal, 75775 PARIS CEDEX 16 - No. 42339  1982
PRINTED IN FRANCE
(59 82 03 1) ISBN 92-64-12365-2